The Historical Development of Chemical Concepts

Chemists and Chemistry

VOLUME 12

A series of books devoted to the examination of the history and development of chemistry from its early emergence as a separate discipline to the present day.
The series will describe the personalities, processes, theoretical and technical advances which have shaped our current understanding of chemical science.

ROMAN MIERZECKI
*Faculty of Chemistry, University of Warsaw,
Poland*

The Historical Development of Chemical Concepts

Kluwer Academic Publishers
DORDRECHT/BOSTON/LONDON

PWN — Polish Scientific Publishers
WARSZAWA

Library of Congress Cataloging-in-Publication Data
Mierzecki, Roman.
  [Historyczny rozwój pojęć chemicznych. English]
  The historical development of chemical concepts / Roman Mierzecki
; [translated by Andrzej Diniejko].
     p.   cm.—(Chemists and chemistry: v. 12)
  Translation of: Historyczny rozwój pojęć chemicznych.
  Includes bibliographical references (p.) and index.
  ISBN 0-7923-0915-4
  1. Chemistry—History.   I. Title.   II. Series.
QD11.M6613 1990
509—dc20   90-5256

Revised and enlarged translation from the Polish original .
Historyczny rozwój pojęć chemicznych
published in 1985 by Państwowe Wydawnictwo Naukowe, Warszawa

Translated by *Andrzej Diniejko*

Published by PWN—Polish Scientific Publishers,
Miodowa 10, 00-251 Warszawa, Poland
in co-edition with Kluwer Academic Publishers,
P.O. Box 17, 3300 AA Dordrecht, The Netherlands

Distributors for the U.S.A. and Canada:
Kluwer Academic Publishers,
101 Philip Drive, Norwell, MA 02061, U.S.A.

Distributors for Albania, Bulgaria, Chinese People's Republic, Cuba, Czecho-
Slovakia, Hungary, Korean People's Republic, Mongolia, Poland, Romania,
the U.S.S.R., Vietnam, and Yugoslavia:
ARS POLONA Krakowskie Przedmieście 7, 00-068 Warszawa, Poland

Distributors for all remaining countries:
Kluwer Academic Publishers Group,
P.O. Box 322, 3300 AH Dordrecht, The Netherlands

All Rights Reserved
Copyright © 1991 by PWN—Polish Scientific Publishers—Warszawa
No part of the material protected by this copyright notice may be reproduced
or utilized in any form or by any means, electronic or mechanical, including
photocopying, recording, or by any information storage and retrieval system,
without written permission from the copyright owner.

Printed in Poland by D.N.T.

To the memory of my wife
ANNA MIERZECKA

Contents

Preface IX

Chapter 1. Division of the History of Science 1
  1.1. The Period of Practical Arts 1
  1.2. The Period of Greek Science. Antiquity 2
  1.3. The Middle Ages 7
  1.4. The Period of Quantitative Research 16
  1.5. The 19th Century 21
  1.6. The 20th Century 23

Chapter 2. The Element 28
  2.1. Antiquity 28
  2.2. Alchemy 36
  2.3. Weight Methods and the Concept of the Element . . . 48
  2.4. The Theory of Phlogiston 63
  2.5. The Discovery of Gases 69
  2.6. Overthrow of the Phlogiston Theory and of the Aristotelian Elements 72
  2.7. The Foundations of Chemistry according to Lavoisier . 78
  2.8. The Imponderable Elements 81
  2.9. Significance of the Term 'Element' in the 19th and 20th Centuries 90

Chapter 3. The Elementary Particle of Matter 93
  3.1. The Conceptions of Continuity and Discontinuity of Matter in Antiquity 93
  3.2. Rebirth of the Corpuscular Theory in the 14th–18th Centuries 97
  3.3. The First Quantitative Laws in Chemistry 106
  3.4. Corpuscular Views of Dalton 109
  3.5. Atomistic Views of Dalton 116
  3.6. Views on the Divisibility of the Atom in the 19th Century. Chemical Equivalents 117

CONTENTS

3.7. The Fate of Avogadro's Hypothesis 131
3.8. Classification of the Elements 135
3.9. Energetism Against the Corpuscular Theory 145
3.10. Stoichiometric and Non-stoichiometric Compounds . . . 149
3.11. Parts of the Atom 151

Chapter 4. The Structure of Chemical Compounds 159
4.1. Structure in Inorganic Chemistry 159
4.2. The First Theories of the Structure of Organic Compounds 168
4.3. Valency 175
4.4. Rational Formulae 179
4.5. The Structure of Unsaturated and Aromatic Compounds 183
4.6. Stereochemistry 186
4.7. Conformational Analysis and Rotational Isomerism . . . 191
4.8. The Theory of Ions 194
4.9. Extension of the Concept of Valency: Complex Compounds 196
4.10. Electrons and the Chemical Bond 199

Chapter 5. Capacity of a Substance for Transformation 207
5.1. Antiquity 207
5.2. Chemistry in the 11th–18th Centuries 211
5.3. Boyle's Corpuscular Approach 213
5.4. Affinity in the 18th Century 215
5.5. Heat As a Measure of Affinity 220
5.6. Formation of Substances in Living Organisms 224
5.7. Chemical Equilibrium and Reaction Rate 229
5.8. Is it Really the Heat of Reactions? 235
5.9. Catalysis and Activation Energy 241
5.10. Irreversible Thermodynamic Reactions 243

Postscript: a Reflection 246
Chronological Tables 247
References 261
Index of Names 270
Index of Subjects 276

Preface

Since the dawn of history man has been trying to understand the physical world. For that purpose he has created concepts which help him to describe his observations and formulate conclusions. As he has accumulated information, the concepts have undergone transformations, and frequently the same term refers to gradually changing notions. The aim of this book is to trace, within the compass of the past centuries, changes in the content of a few fundamental concepts which are essential for the understanding of chemical phenomena. These concepts include: the element, the elementary particle of matter, the structure of chemical compounds, and the capacity for transformation. These problems are so fundamental that the work must necessarily embrace a significant part of the history of chemistry.

In considering the above concepts, one should always bear in mind that they have been conceived by man, so their ideal representations will not be found in physical reality. The meanings of particular concepts are not sharply delineated; they frequently overlap because they describe one and only one nature. They embrace not only what we call chemical phenomena, but sometimes also physical or biological phenomena, since there exist no confines between man-made concepts and between the particular branches of knowledge. The division of knowledge into particular sciences serves only to introduce some systematization into the information attained as well as to the methods of study or practice.

Textbooks on the history of chemistry are usually concerned with problems, chronology, and biographies. In particular textbooks each of these topics plays a more or less dominant role. In the present book problems are emphasized. I wish to tell the reader how chemistry has attained its present-day state of development. I begin my discussion of each of the above concepts with ancient times. Scientific argumentation in those times was entirely different from ours, but formulations made by the ancients became a basis for the present-day notions.

Therefore, a relatively large amount of space will be devoted to the old concepts and a number of original texts will be quoted. The closer we come to our modern times, the easier it will be for us to understand the views of particular investigators, and the more often it will suffice merely to summarize their opinions. Therefore, the revolutionary development of chemistry in the last fifty years is mentioned only cursorily. Students of chemistry know this period well from textbooks and lectures; and even the most concise but lucid presentation of this period would greatly exceed the space provided for this work. I hope that the reader will notice that each of the old theories—no matter how absurd it might seem today—was quite plausible if we view it against the background of contemporary science; each concept was intended to account for concrete observations.

As the development of each chemical concept will be discussed in a separate chapter; it stands to reason that at the outset the history of chemistry should be shown in relation to the general advancement of science. Chronological tables at the end of the book will help the reader to trace the history of chemistry throughout the ages.

So many investigators have contributed to the development of chemistry that no textbook on the history of science can mention them all. In the present book the development of chemical concepts is illustrated by the opinions of some chemists whose approach was characteristic of their times, or—conversely—chemists who overthrew prevailing opinions. Extensive quotations of original texts are provided. I believe that this will be a more interesting way of presenting new theories and opinions, exactly as they were formulated by their authors, rather than to comment on them. Commentary always distorts the original ideas; and the commentator, adaptating them to his didactic purposes, frequently filters new ideas through the contemporary state of the art and consequently overlooks many of their innovative values.

The ever-increasing amount of information which should be conveyed to science students during lectures in a concise and systematic way has gradually forced historical themes out of university curricula. However, in recent years in both Western and Eastern Europe an awareness of the importance of the history of science has grown anew. Such an opinion was formulated by C. A. Russell in the introduction to the book *Recent Developments in the History of Chemistry* published by The Royal Society of Chemistry in 1985 [17]; it can also be found in B. Wojtkowiak's textbook entitled *Histoire de la chimie* [22] as well as in the work of three German historians of chemistry, I. Strube, R. Stolz and H. Remane, entitled *Geschichte der Chemie* [20]. This last publication is a textbook designed for students taking an obligatory course of the history of chemistry at faculties of chemistry in the German Democratic Republic. Attempts to introduce a course on the history of chemistry are also being made in Czechoslovakia, Poland, and the USSR.

In the opinion of many lecturers, courses in the history of particular sciences should be offered for undergraduates who have already studied the discipline of their science to a degree that its history may be presented in the light of the present state of the art. Such an approach will enable them better to understand the development of a particular science and the meaning of currently used concepts rather than selective and random references made during other courses. Therefore, it seems that university lectures on the history of science should present the state of the art of a given science and the debate held in the successive periods, and within this framework the development of particular ideas can be discussed. However, a presentation in the form of a book may be based on the discussion of the development of particular problems of science.

Roman Mierzecki

CHAPTER 1

Division of the History of Science

1.1. The Period of Practical Arts

Natural science has grown from observation. From the dawn of history man has watched nature and handed down the results of his observations to the following generations. At the outset he utilized various natural substances, and subsequently he learnt to process them. In such a way different arts emerged, some of which later developed into chemistry. They chiefly included those which required the application of fire, i.e. metallurgy, pottery, the preparation of food by heating as well as husbandry and dyeing. This list of practical crafts was further extended by the search for and processing of substances of personal adornment, such as pigments and perfumery extracts. Husbandry was based on reproducing nature. The arts which made use of fire enabled ancient men to manufacture qualitatively new products through physical and chemical transformations.

Metallurgy has had a particular significance in the development of civilization. At first man found metals in the pure form only in meteorites which miraculously—as he thought—fell from the sky to the Earth; later he learnt to smelt metals from ores. The significance of metallurgy is reflected in the modern names of the periods in human culture: the Bronze and Iron Ages. Hard tin bronzes which melted at lower temperatures than pure copper were produced in Egypt, Mesopotamia and Crete as early as 3000 years BC. Bronze smelting and casting were known in Southern Europe about 2000 years BC, and in Poland about 1700 years BC. The spread of iron in the second millennium BC gave rise to the so-called Iron Age. Admittedly, objects made from meteoritic iron in the 4th–3th millennia BC have been found in Egypt and Western Asia, but it was not until *c.* 1200 BC that the skill of smelting iron was acquired in Asia Minor and Egypt; *c.* 1000 BC in southern Europe. The Iron Age lasted until medieval times, i.e. the 13th century.

The appearance and formation of practical arts can be traced to prehistoric times and therefore this period in the development of science is called the *period of practical arts*. It was characterized by accumulation and transmission of experience necessary to obtain certain products. It was not known from the start, for example, which material is most suitable for moulding a pot, how it should be moistened, how and at what temperature it should be fired, and finally how it should be glazed so that water might not leak through its walls. This transmission of information was not, however, accompanied by attempts to generalize acquired experience, which is an inherent characteristic of any scientific activity.

1.2. The Period of Greek Science. Antiquity

The first attempts in the history of mankind to generalize observations were made in the 6th century BC. For European culture the considerations of some philosophers who lived among the Greek tribes on the western coasts and islands of present-day Turkey and in the Peloponnese were of particular significance. They initiated what we call the *period of Greek science* which lasted two thousand years until the 17th century. At the onset of the development of scientific concepts, in the 6th century BC, two fundamentally different views on the nature of the physical universe were formulated. The philosophers from the Ionian Isles who called themselves *naturalists* (Greek *fisikoi*) relied on the observation of nature and sought generalizations concerning the origin and transformation of the universe, and rejected mythological interpretations. Thales of Miletus is considered to be the most outstanding representative of the Ionian philosophers. A little later (at the turn of the 5th century BC), the Ionian philosopher Heraclitus emphasized the significance of changes occurring in the environment and he is said to have enunciated his famous *panta rhei* (everything is continually in flux), which illustrates his view of the natural phenomena. The Ionian philosophers sought the laws which govern the transformation of nature.

1.2. THE PERIOD OF GREEK SCIENCE. ANTIQUITY

The second group consisted of philosophers who were active in the town of Elea, and who were hence referred to as the Eleatics. They held that truth could only be discovered by reasoning. In their view the world was one and unchangeable. One of them, Parmenides is credited with saying that "Things are neither born, nor are they dead; this is only a deceptive appearance". This statement may be regarded as one of the first formulations of the laws of conservation which are essential in modern natural science. It was later rephrased by Aristotle in his treatise *On the Heavens*: "It is impossible for a thing always to exist and yet to be destructible" [23, 282a].

Likewise, in the didactic poem of Lucretius *De rerum natura*. (On the Nature of Things) from the first century BC, this law is expressed in the following form: *Nec stipata magis fuit umquam materiai copia nec porro maioribus intervalis, nam neque adaugescit quicquam neque deperit inde*. (The matter of the whole universe never was either more or less condensed than it is now [24].)*

The law of conservation of matter, formulated throughout the centuries in a variety of ways, has never in fact been contested. This problem will be elaborated later on in the book. Let us note, however, that Greek word *yle*, or Latin *materia*, means substratum, a physical substance and the Greek *physis* means nature.

The two approaches to the natural sciences discussed earlier weighed heavily on the whole period of Greek science, and particularly on its development in ancient Greece and Rome during the first centuries of the modern era. Of course, neither the Eleatics, who emphasized reasoning, nor the Ionian philosophers, who observed nature, performed any experiments. When the latter spoke about 'experiment', they meant 'observation'. From their first experiments they drew definite conclusions, and in further 'observations', which included both inanimate and animate nature, and even mental phenomena, they sought confirmation of those conclusions (as late as the 17th century, Robert Boyle criticized the manner of arbitrary selection of demonstrations

---

* The English translator adds here a commentary: "... as his atoms are eternal, it is an axiom that none come into being or go out of being; the sum of matter therefore must ever be the same".

in order to prove the previously adopted assumptions). The slave system of the ancient world inculcated in people a contempt for manual work, including experimentation, whereas it favoured the rational analysis of observations. In consequence, several methods of reasoning were then devised and through continuous development they have survived until the present time.

Of paramount importance for the description of observed phenomena was an approach developed by philosophers led by Pythagoras. They discovered that vibrating strings produce harmonious notes only when their lengths are in fixed relationships, with one another, expressed by small integers. On this basis the Pythagorean philosophers conceived that the harmony of the world, and the laws which govern it, could be described by means of numbers. Thus the Pythagoreans introduced mathematics to the description of natural phenomena. Such a figurative representation necessitated some degree of idealization of phenomena; and let us note that the Greek term *idea* denotes image.

The idealistic approach to matter was later developed by Plato who held that in reality there exist only ideas of objects, feelings and sensations, which have ideal forms. Concrete objects and sensations are accessible to man only as imperfect reflections of their ideas–designs. Thus, in order to understand reality, one must reach those pure ideas through reasoning. This view, which emphasizes the real existence of ideas, is referred to as *objective idealism*. Plato's idealism with its variations exerted a great influence on the first Christian philosophers and it would pass into science until the 12th century with enormous effect. Its traces can be seen to the present day in the form of the necessary idealization which accompanies the formulation of new scientific concepts, as well as in a frequently futile search in nature for the representation of man-made models.

Such idealization was rejected by Plato's pupil, Aristotle, the first great classifier and collector of data in science. The starting point of his enquiry was experience, which meant observation of concrete phenomena actually occurring in nature. His reasoning consisted of two stages: a stage of drawing general conclusions from observation (induction), and a stage of drawing detailed conclusions from general conclusions (deduction). Detailed conclusions referred to the causes of

1.2. THE PERIOD OF GREEK SCIENCE. ANTIQUITY

particular phenomena; the term 'cause' had then a wider connotation than it has today. Aristotle held that there exist four casual determinants of each phenomenon:

1. Material determinant, which means the material out of which an object is generated and which is immanent in the generated object.

2. Formal determinant may mean the form or pattern, i.e. what the thing is defined as being essentially; and also the genus to which this essence belongs.

3. Efficient determinant is the immediate cause of change or of cessation from change.

4. Telic determinant may mean the end (*telos*) or purpose for the sake of which a thing is done.

The distinction of the last kind of determinants strengthened the view, held for many centuries, that each phenomenon must have a concrete end or purpose. This view, referred to as *teleology* prevailed in science until the 18th century, and its traces can still be found in some disciplines. That kind of reasoning determined Aristotle's views on the texture of matter which are discussed in Chapter 3. Aristotle taught his pupils while walking, hence his pupils and adherents of Aristotelianism were called *peripatetics* (Greek *peripatetikos*—from *peripatein*—to walk up and down). Democritus used another kind of reasoning. He accounted for all observable qualitative changes by quantitative changes of invisible particles.

A different approach can be found in the interpretations of the movement of celestial bodies made by some Greek philosophers. They imagined that those bodies orbited on the surfaces of the spheres and various philosophers fixed the centres of the spheres at various points of the Universe. As a matter of fact, they built models in order to reflect the observable reality. In the second century, Ptolemy appears to have proposed the most compatible model of epicycles.

We should also recall Euclid who, in the fourth century BC, based geometrical considerations on a system of axioms and created the original axiomatic reasoning.

The ideas and the kind of reasoning formulated by the Greek philosophers played a significant role in the formation of the whole of European civilization; however, they did not contribute to the development of the practical arts in antiquity. The overall level of the individual arts did not differ much from that in the preceding period. Most of the population seems to have been quite satisfied with available raw materials and the simple methods of processing them, based on slave labour. However, this situation gradually changed as the centuries went by. At the end of the ancient times the population had grown so much that the methods of primitive artisans proved to be inadequate. New techniques were sought for the production of necessary materials, as well as the imitation of some of them. This finds evidence in recipes on papyri found in Egypt and dating from c. 250. They are now kept in the museums of Stockholm and Leyden. These recipes are chiefly concerned with the following arts: dyeing, mining of gem-like minerals, and producing bronzes and alloys which resemble silver and gold; they also provide ways of covering objects with a layer of silver or gold, or with a layer which would imitate these metals. Such an effects was obtained, for example, by covering an object with an amalgam of gold or silver and allowing the mercury to evaporate. Gilding was also imitated by applying to an object a coat of paint containing saffron and gall. As can be seen such exertions were frequently aimed at producing a kind of substitute for some substances, and not at the actual transformation of one substance into another.

In the 3rd and 4th centuries of the modern era the Library and Museum in Egyptian-Greek Alexandria was the main centre for scientific research of the ancient world. We find the term *chymea* for the first time in the writings of the Alexandrian scholar Zosimos. The origin of this term is not well known. According to one hypothesis, it was probably derived from the name *chemi* which referred to Lower (Black) Egypt, a country which was undoubtedly a cradle of chemistry thanks to the development of embalming methods. The Egyptian god Toth identified with the Greek god Hermes* is reputed to have originated the chemical art.

Another hypothesis associates the term 'Chemistry' with the Greek word *chymea* which denoted an art of smelting metals or with the word

---

* Hence chemistry was called a hermetic science in the Middle Ages. This term in another sense has been used until today.

chymos which meant 'juice' flowing from treated ores. There is also another hypothesis which claims that the term 'chemistry' stems from the Hebrew core *chm* meaning something hot.

1.3. The Middle Ages

At the turn of antiquity and in the mediaeval period, a significant impact on the development of science was exerted by the decline of the western Roman Empire and an increase in power of the Arabian states. Christian culture grew on the ruins of the Roman Empire; however, the main centres of advanced study were situated in the territories where Arabic culture dominated. Early Christian culture lost almost all contacts with the cultural heritage of ancient Greece, only Plato's conceptions remaining commonly known and recognized. The Arabian scholars translated into Arabic many Greek works, and Arab as well as Jewish philoshers, who were active within their sphere of influence, developed Aristotle's theses in their commentaries. The most outstanding philosophers of that time included: Jabir-ibn-Haiyan better known as Geber, who lived in the 8th century and translated many Greek works, and the Arab philosopher and physician Abu Ibn Sina better known by his Latin name Avicenna, who lived at the turn of the 10th century. The Arabs added the Arabic article *al* to the Zosimosian term *chymea*, hence a new term 'alchemy' was coined, and adherents of this new art were called 'alchemists': Arabic culture as well as Arabic philosophy were characterized by a holistic approach to emotional and natural phenomena. The Arabic alchemists treated natural science as a whole without making any distinction between inanimate and animate nature, including a separate science about man as well. There was a close link between what we now call chemical and medical considerations. Particularly, they advanced the science of the Greek philosopher Galen about the use of vegetable products as drugs.

It was not until the 12th century, in the period of the fall of the Arabian Empire, that Christian philosophers took an interest in the works of Aristotle which they found in Arabian libraries. Numerous

Latin translations of the Arabic versions of the works of Greek philosophers were then made by monks. Adelhard of Bath, who was active at Chartres, revived the conceptions of Democritus. Albert of Bollstadt (Albertus Magnus) and Thomas of Aquine, who was active in the early 13th century, attempted to reconcile the ideas of Aristotle with Christianity. Aristotle was also translated at Oxford where Robert Grosseteste and his pupil Roger Bacon were the foremost scholars of the period. The philosophers of the Oxford school pondered over the conditions which must be fulfilled so that correct inferences might be reached by the inductive method and false inferences avoided. A particular role of systematic observations was emphasized as a method of checking the truth of inferences. Roger Bacon asserts:

Experimental science has three great prerogatives over all other sciences: it verifies conclusions by direct experiment; it discovers truths which they could never reach; and it investigates the secrets of Nature and opens to us a knowledge of the past and of the future [25].

He also considered the possibility of a practical application of the acquired natural science. The approach of the Oxford scholars to the observation of nature reveals some influence of the Pythagorean school. They tried to express their observations in a mathematical or, more precisely, a geometrical form because, in the 13th century, geometry was the best developed branch of mathematics. Thus, the first attempts at a quantitative approach to the observed phenomena were made as early as the 13th century. Four centuries later this approach became a dominant method of scientific research.

Some authors point out that the development of empirical methods which opposed pure speculative philosophizing contributed to the reconciliation, in some measure, of Christian philosophy with the contested pagan philosophy of ancient Greece [12].

Of the numerous scholars who developed the Aristotelian analysis of observation, mention should be made of William of Occam (14th c.) who endeavoured to simplify the derived inferences. His statement known as 'Occam's razor', *Entia non sunt multiplicanda praeter necessitatem* (Entities should not be multiplied unnecessarily) emphasized that it is reasonable to minimize the number of concepts and do away with concepts which do not contain anything new, but which duplicate other concepts.

Discussing the advancement of science in the late Middle Ages, it is important to see the external factors which enabled science to be developed. In the 11th century, translators and commentators of the works of Greek philosophers were mainly concentrated in ecclesiastical centres, such as Chartres or Bologna (a school of legal studies). Those communities gave birth to universities, i.e. the organizations of teachers and students engaged in the pursuit of learning. The first university was created in Paris in *c.* 1160. In the year 1167, some of the lecturers

Fig. 1. Alchemical laboratory,
from *Liber Alze* in *Dyas Chimica Tripartita*, 1625.

of Paris university founded a university or, more precisely, only one college at Oxford. Robert Grosseteste became its first rector. In 1209, Cambridge University was created as a branch of Oxford University. In the 13th and 14th centuries, many other universities were founded; amongst others in Padua (1222), Naples (1224), Prague (1347), Cracow (1364), and Vienna (1467).

Although mediaeval chemistry is usually referred to as *alchemy*, the latter included not only chemistry but also other natural sciences and lasted much longer than the Middle Ages. However, the achievements of practical chemistry in the Middle Ages should be separated from the alchemists' outlook. Of course, the mediaeval alchemist-philosophers were frequently good chemical practitioners. Bolos of Mende, who lived in the 2nd century BC, is reputed to have been the first alchemist, and Johann Wolfgang Goethe, who lived at the turn of the 18th century, is believed to have been the last. The well-known Romanian ethnologist Mircea Eliade has traced the roots of the alchemical outlook in the consciousness of the primitive peoples from the period of the practical arts [26].

Alchemical writings first circulated among Alexandrian, and subsequently among Arab scholars and, from the 13th century, also among European scholars. The greatest number of such writings was published in the 17th century after the spread of printing [27, p. 112].

A major achievement of practical chemistry in the Middle Ages was the development of the techniques of separation and purification of substances. Distillation had already been known by the end of ancient times. Different kinds of crystallization, sublimation, and different methods of heating were discovered in the Middle Ages. In consequence, chemistry was then called a *spagyric* or *hermetic-spagyric* science (from Greek *spao*—divide, *ageiro*—bind). Separation methods enabled some new substances to be isolated: first of all the main inorganic acids and alkalies, as well as antimony and phosphorus, which are nowadays regarded as elements. Acids were obtained by the roasting of salt mixtures and liquefaction of the liberated vapours, which were believed to be the spirits of those salts. For example, in order to obtain *aqua regia*, which dissolves the king of metals (gold), a mixture of saltpetre and sal ammoniac was roasted [28]. Alkalis were produced by leaching the ashes of certain plants. Practising chemists were also engaged in the production of different alloys. The observed transmutations of substances had a great significance for alchemical interpretations.

The alchemist's outlook was based on entirely different foundations than that of the modern scientist's. Today preference is given to an analytical approach to natural phenomena, which is in turn necessary to permit syntheses. The reasoning of the alchemists was—as we have

1.3. THE MIDDLE AGES

noted—holistic. They assumed that all transformations occurring in both inanimate and animate nature, in man and among the permanent and wandering stars, were interrelated and exerted an influence upon each other. What is happening in the macrocosm must also be happening, the alchemists believed, in the microcosm. They included man's mental life in the sphere of the microcosm. All substances, and man, are striving spontaneously towards improvement. However, the rate of natural transformation is different; it is much slower in inanimate than it is in animate nature. The alchemists thought that their task was to quicken the rate of natural transformations, which improved base metals and minerals and turned them even to the most noble metal—gold. The alchemists showed the utmost reverence for the materials and utensils they were using and a deep devotion to their practices.

In the opinion of the alchemists the union of contrarieties was the essential factor which initiated all transformation and development. This view originated in prehistoric times and played an important role in antiquity. Union of contrarieties was most overtly manifested in the act of copulation when the male and female elements were united. The alchemists sought the same kind of bound in every creation and procreation. For the earth to yield crops, it must first be fecundated by the blade of the plough; and in 1616 the German alchemist Michael Maier wrote:

Nam materia omnis, ut foemina, per se frigida et ad generationem inepta est, nisi excitetur calore et motu suae formae, tanquam agentis masculi aquo ut initium generationis capit, sic perfectionem et finem nanciscitur [29].
(The whole matter, like a female is in itself frigid and unable to give progeny unless its form is stirred by heat and motion of the male agent whereby fertilization occurs in order to proliferate perfectly and finally.)

The alchemists believed that, like plants on the earth, ores and metals grow and strive towards improvement deep under its surface. The improvement of metals which the alchemists endeavoured to achieve in various chemical operations was meant rather to speed up the natural process, and not to cause an artificial transformation. Particular metals were not regarded as simple substances, and alloys were thought to be new metals. Obtaining alloys having properties approaching those of silver or gold supported the belief in the transformation of base metals into noble ones.

Curing sick people was believed to be an analogous process to the improvement of metals. In order to improve base metals and to cure a sick person the alchemists sought a philosopher's stone which was believed to have the power of transmuting base metals into gold and to be a remedy for all ills. The philosopher's stone was to be on one hand some kind of sophisticated white powder produced artificially; but on the other hand it could be regarded as a symbol of the alchemist's skill and art. In the work *Rosarium Philosophorum*, written by Arnold de Villanova in the thirteenth century, the following statement can be found: "He who can make it knows what it is like" and Zosimos asserts: "This unknown is known to everybody" [27]. Some philosophers attribute to the philosopher's stone a mystical significance or the idea of the supreme initiation [27, p. 29].

Thus alchemy included all contemporary sciences, and the outstanding role of faith associated with alchemy distinguished it from modern natural sciences. No wonder that chemistry which belonged to alchemy was closely linked with medicine. In the 16th century medical chemistry (called *iatro-chymia* or *chym-iatria*) emerged as a predecessor of pharmacy.

Its creator and main representative was a Swiss alchemist Theophrastus Bombast von Hohenheim who assumed the pseudonym Paracelsus. He introduced mineral substances to medical care.

The seventeenth century textbooks of alchemy chiefly contain recipes for preparing ointments, infusions, and other medications as well as descriptions of how to dissolve gold in mercury, how to recover it, and how to obtain alloys. They also contain recipes on how to obtain the philosopher's stone. The latter problem was also dealt with in a treatise on alchemy written in 1586 by the most outstanding Polish alchemist Sędziwój (Sendigovius), in which the author proposes twelve 'operations' for producing that miraculous agent. Yet none of the products of those operations had the desired properties [30, pp. 95–127].

Recognizing the unity of nature and interaction of all processes, the alchemists believed that by making and observing chemical transmutations they participated in them too. And since such transmutations were aimed at the ennoblement of matter they felt ennobled, too. They repeated the same reactions again and again in order, not to obtain the already well-known results, but to make themselves spiritually ennobled. The alchemists thought that a transmutation could be successful only if they were in a state of grace. Therefore, in many chemical laboratories, small chapels could be found, and the alchemists would always begin their practices with prayers. And as they seemed to believe that all phenomena were interrelated, the time of day and the arrangement of the planets exerted an influence of the results of transmutations. This was also closely linked with views on the origin of metals, which will be discussed in detail in the next chapter.

It should be remembered, however, that alchemists did not consider their art to be magic; they did not claim to have used supernatural forces. Paracelsus, whom we have mentioned earlier, defined alchemy as "a way of preparing natural substances", and in the first comprehensive textbook called *Alchemia* written by the German alchemist Andreas Libavius (Libau), we find the following definition: "Alchemy is an art of improving properties and obtaining pure essence from compound bodies by separation" [31].

Medieval chemical texts which describe the technical side of transmutation procedures are written quite clearly, although the terminology used differs significantly from that we use nowadays. On the other hand, alchemical texts which attempted to account for the observed phenomena are written in a hermetic style and contain symbols which were intelligible only to the adepts. The symbolism of alchemical writings was manifested in their language as well as in graphical signs and figures. In order to illustrate this, we shall cite some alchemical texts in subsequent chapters. Now we shall present one of the oldest and most famous symbols—*ouroboros*: a devourer of its tail represented in the shape of a snake, because a snake sloughs its skin (Fig. 2). It was most often treated as the symbol of mercury—the basic substance. From available ancient and alchemical writings, beginning with the Sumerian texts, it follows that this symbol had an extremely wide meaning. It not only conveyed an idea that primary matter can create all substances, but it also represented all changes that occur

14   1. DIVISION OF THE HISTORY OF SCIENCE

Fig. 2. Ouroboros—the symbol of mercury, from *Lambspring* in *Dyas Chimica Tripartita,* 1625.

cyclically, i.e. revolutions of the wandering stars (the Sun, the Moon, and the planets), and seasonal changes. Besides, it was believed that the snake rejuvenated itself after throwing off its skin; hence ouroboros was the symbol of the eternal human dream of preserving permanent youth.

Changes, chiefly changes of colour when a substance was subject to the effect of fire, were symbolized by the salamander. We have already mentioned that a juxtaposition of contraries played a significant part in the alchemical outlook; we shall return to this problem in the next chapters. The interacting contraries were represented as two birds—a white and black one which devoured each other. It was also believed that the contraries are manifested in man as soul (Latin *anima*) and spirit (Latin *spiritus*) which occupy the common body. A frequently represented illustration of this notion is a deer with rich antlers (soul), and a unicorn (spirit), living in the common wood (body) (Fig. 3). Let us note that the opposition between body and soul, which exists to the present day in Christian philosophy, had its origin with the Greek philosophers. For Anaximenes, breath (Greek *pneuma*) was a sign of

Fig. 3. Anima (deer) and spiritus (unicorn) in a body (forest); a symbolic representation, from *Lambspring* in *Dyas Chimica Tripartita*, 1625.

life—the most essential element in nature (cf. Chapter 2). This distinction was later confirmed by the observations of mediaeval technological processes, particularly distillation, during which a new substance was obtained as a result of the liquefaction of 'spirits' or vapours released by a heated mixture.

The symbolic style of the contemporary chemists was ridiculed by Robert Boyle in his work *The Sceptical Chymist* published in 1661:

(...) chymists write thus darkly, not because they think their notions too precious to be explained, but because they fear that if they were explained, men would disern, that they are farr from being precious [32, p. 114].

It is interesting to note that in his writings Boyle does not use the term 'alchemists', but he criticizes the contemporary 'chemists'.

1.4. The Period of Quantitative Research

As we have said earlier, from the late 13th century mediaeval scientists began to verify rational inferences in observations and then represented the results obtained in a geometrical form. Gradually, more and more importance was attached to that kind of analysis of phenomena, as quantitative description was found to be more valuable than qualitative. In astronomy this new approach was reflected in Copernicus' views published in the mid-15th century who, on the basis of the observation of the movements of the moon and mathematical calculations, proved that the heliocentric description of the movement of the planets is simpler than the geocentric model. The spherical trajectories of the planets assumed by Copernicus were in agreement with Plato's theories of movement along ideal circles.

The application of mathematical methods to the natural sciences, which was begun in the 13th century and was generally widespread at the turn of the 16th century, is regarded as a manifestation of a decisive break-through in the scientific approach and is called a revolution by some historians of science. Therefore, the 16th century is regarded as the end of the period of Greek science, and the 17th century marks the beginning of the period of quantitative research. In physics, the revolution in scientific thought is associated with the name

1.4. THE PERIOD OF QUANTITATIVE RESEARCH

of Galileo, whose work is the best manifestation of a new approach to research. The development of the mathematical description of phenomena strengthened the role of experiment in its new form. As we have mentioned earlier, in the Middle Ages, the term 'experiment' meant the observation of phenomena under natural conditions. Galileo, however, deliberately changed the initial conditions in order to observe the quantitative results of the introduced change.

On the basis of such experiments on the falling of bodies he presented the following argumentation in 1590:

(...) all compound and non-compound bodies have greater or smaller weights (...), thus we say that fire is lighter than air not because it has no weight, but it has a smaller weight than air (...) and if air were removed from under fire, who would doubt that fire would take the place of air? And as there is nothing in a vacuum, what can be less heavy than nothing? [33, p. 29].

Experiments with a barometric tube conducted by Torricelli and Pascal in 1643 and 1648, respectively, proved directly that air has weight.

Thus Galileo drew conclusions from purposefully planned experiments. In this way he developed the reasoning of the ancient Pythagoreans, assuming that systematic experiments could disclose the mathematical harmony of the universe. With this end in view he employed a hypothetico-inductive method which was complemented by idealization, i.e. he adopted the prevailing property as an hypothesis and discarded less significant properties. For example, Galileo inferred that all bodies should fall in a vacuum at an equal rate irrespective of their mass, although he was unable to obtain a vacuum in order to verify this fact experimentally. However, he realized that a side-effect—air resistance—caused a deviation from the formulated law.

The Aristotelian inductive-deductive scheme of reasoning was also critically analysed by Francis Bacon. He stated that not every induction must necessarily lead to correct inferences. In the case where several conclusions may be drawn, contradictory to one another, Bacon proposed the performance of an experiment which would eliminate all inferences but one (the method called in Latin *experimentum crucis*). In his works he emphasized that science should aim at the development of its practical applications, and also claimed that science has nothing in common with theology and teleology. In this respect he developed the views of his 13th-century compatriots—Robert Grosseteste and Roger Bacon.

In the next century Isaac Newton also analysed the inductive-deductive method. He pointed out the great significance of deductive inferences going beyond premises which form the basis of an inductive stage, but such inferences have to be confirmed by experiment. Furthermore, Newton developed an axiomatic approach to the analysis of phenomena. He distinguished three stages of the axiomatic method: (1) formulation of a set of axioms; (2) presenting a method of coordinating the set of axioms with experimental data; (3) experimental confirmation of inferences derived from the set of axioms [13].

The development of quantitative methods favoured the spread of the corpuscular view of matter. Both Boyle and Newton supported the idea of the smallest particles of matter; the latter named them 'ultimate particles' (cf. Chapter 3).

Throughout the seventeenth century quantitative methods find an increasingly wider application in chemistry. Initially they are weight methods which have been used since mediaeval times to determine the alloy composition. In the seventeenth century the weight methods were employed by Rey, and next by Boyle, but it was Lavoisier who used a balance most consistently late in the eighteenth century.

Early in the eighteenth century, when quantitative methods were already being spread, the first chemical theory—the 'theory of phlogiston'—was formulated; it enabled the observed chemical reactions to be arranged in some order. Although it had its origin in the period of quantitative measurements, it should be stressed that it was a qualitative theory which included very many elements of the preceding period of Greek science. It was, however, the first chemical theory which formed a basis for the development of new technological processes (cf. Chapter 2).

An important manifestation of the ever increasing role of the quantitative approach was formulation of the quantitative laws in chemistry at the turn of the 18th century. The first one was the so-called 'law of conservation of mass' which was never contested but

1.4. THE PERIOD OF QUANTITATIVE RESEARCH

which was not verified until the early 20th century. On account of its exceptional significance, we shall discuss its origin.

As we have said earlier, in antiquity Parmenides and later the Roman writer Lucretius formulated the principle of the conservation of matter. It will be difficult to mention all the authors who proclaimed this principle in a variety of ways; therefore, we shall quote only a few formulations from the seventeenth and eighteenth centuries. In his work 'The Sceptical Chymist' Boyle states:

(...) for it far exceeds the power of merly natural agents, and consequently of the fire, to produce anew, so much as one atome of matter, which they can but modifie and alter, not create; which is so obvious a truth, that allmost all sects of philosophers have denied the power of producing matter to second cause [32, p. 63].

We can presume that Boyle—in accordance with the contemporary views—regarded God as the first cause of the formation of matter. A similar opinion, expressed by Newton in 1717, will be presented in Chapter 3. Now we shall point out that in one of the first Polish textbooks of physics published in 1765, Father Józef Rogaliński, 'a praeceptor of mathematics and experimental physics' in the Jesuit Schools of Poznań, gives the following uncritical account of the views of the Peripatetics:

Primary matter is never first born, but is created by God out of nothing; it never perishes, but when we say a thing is born or decayed, only its form is decayed, but primary matter remains intact and assumes another form [34, p. 9].

A more extensive formulation is provided by the Russian scientist Mikhail Lomonosov in his work entitled *Meditationes de solido et fluido* published by the Academy of Sciences in St. Petersburg in 1760:

Omnes autem, quae in rerum natura contingunt mutationes, ita sunt comparatae, ut si quid alicui rei accedit id alteri derogetur. Sic quantum alcui corpori materiae additur, tantundem decedit alteri. Quot horas somno impendo, totidem vigiliis detraho. Quae naturae lex cum sit universalis, idcirco etiam ad regulas motus expenditur [35, Vol. III, p. 382].

(Thus all transformations occurring in nature are of condition that the more something decreases in one body the more it will increase in another. So, if some amount of matter is added to one body the same amount will be subtracted from another; the more hours I sleep the fewer hours I shall stay awake. It is a general law of nature, therefore it is extended to the laws of motion.)

Identical opinions will be found in a letter which the scientist wrote to Leonard Euler in 1748 and which was published exactly two hundred years later [35, vol. II, p. 282]. Thus Lomonosov formulated a general law of conservation, but like his predecessor he failed to prove its validity.

However, the French chemist Antoine Laurent Lavoisier is commonly credited with having been the author of the law of conservation of mass. A detailed analysis of Lavoisier's texts, particularly his manual entitled *Traité élémentaire de chimie* published in 1789, reveals that he treated the law of conservation of mass rather as a premise and not inference. He made only one passing remark about the so-called law of conservation of mass in Chapter XIII entitled 'On the Decomposition of Vegetable Oxides By Vinous Fermentation'. Lavoisier calculates the amounts of products received and justifies the correctness of the results obtained:

car rien ne se crée ni dans les opérations de l'art ni dans celles de la nature ... [36, p. 141].
(... for nothing is created in the operation of art and of nature, and it can be taken as an axiom that in all operations the quantity of matter before is equal to that found after the operation; that the quantity and quality of the principles (i.e. the elements—R.M.) remain the same; and that there are only changes and modifications.

Upon this principle the whole art of making experiments in chemistry is founded: we must always suppose a true equality between the principles of the body which is examined and those which are obtained on analysis.)

As can be seen, the essential novelty in Lavoisier's formulation is his emphasis on the constancy of the quantity and quality of the particular elements during chemical transformations. We shall show in Chapter 2 that, using this law as a premise, Lavoisier drew quantitative conclusions concerning the complex structure of bodies which had been considered until his time as simple bodies. Let us note that on the basis of one experiment it is not possible to prove simultaneously the validity of both the inference and the premise, just as one cannot obtain the values of two unknowns from one mathematical equation. Thus, Lavoisier did not prove the law of conservation of mass, but since he regarded it as an axiom, he treated it as a premise upon which he built his quantitative inferences; he only gave to this law a new shape, which was particularly applicable in chemistry.

With the growth of science in the 17th and 18th centuries the number of researchers and the intensity of research were constantly increasing. By the 17th century the number of persons engaged in some kind of research had been so great that it became necessary to inform one another about the investigations carried out. This gave rise to the first associations of scholars or societies aimed at the advancement of science.

The Academy of the Lynxes (*Academia dei Lincei*), founded in 1603 in Italy, is still in existence up to the present day. The Invisible College, which has existed in London since 1645, was renamed by Royal Decree in 1660: The Royal Society of London for Improving Natural Science. A few years later l'Academie de Sciences was founded in Paris. The first society of natural scientists on Polish territory was established in Gdańsk in 1743. In the years 1767–1768 the Warsaw Physico-Chemical Society was active for a short time, which edited a journal. Next in 1778, the Warsaw Physical Society was established; however, of greatest importance for Polish scientific life was The Society of the Friends of Science, founded in 1800.

1.5. The 19th Century

In the first half of the 19th century, thanks to the firm establishment of the concept of chemical elements, there occurred—as a result of the consistent application of quantitative methods—an enormous development of inorganic chemistry. Many new compounds were discovered and synthesized. Moreover, a prevailing majority of nineteenth-century scientists held to the corpuscular view of the texture of matter, and thanks to the quantitative chemical laws they formulated, this view began to assume the shape of the atomistic-molecular hypothesis.

In that period a number of new compounds, which are now called aromatic compounds, were isolated from natural materials as well as from the products of coal processing; and methods were devised to introduce some elements and groups of elements into them. Furthermore, it was possible to obtain through synthesis a few chemical compounds which by then had occurred only in living organisms. Thus along with the development of 'mineral' chemistry, the chemistry of carbon compounds, more often referred to as 'organic chemistry', began to evolve in the first half of the 19th century.

In the latter half of the 19th century, the discovery of new compounds and methods became widespread thanks to the rapid growth of chemical industry. The requirements of chemical industry led to a more accurate analysis of chemical processes and a significant improvement in the range and accuracy of measurement. Besides the traditional methods of substance separation, such as distillation and crystallization, as well as gravimetric methods, throughout the 19th century mechanics, electricity, optics, and thermodynamics also became helpful in accounting for chemical processes. This led to the development of 'physical chemistry' as a separate branch of chemistry. The first chair of physical chemistry was established at Leipzig University, Germany, in 1871, and in 1887 a journal *Zeitschrift für physikalische Chemie* commenced publication. The very name 'physical chemistry' was not new. It had already been used in the 17th and 18th centuries in general considerations.

In 1752, Mikhail Lomonosov used the name 'physical chemistry' in his unfinished and unpublished treatise entitled *Prodromus ad veram chymiam physicam*, and in the years 1819–1824 the journal *Annalen der Physik* received a fuller title: *Annalen der Physik und der physikalischen Chemie*.

The quantitative approach applied to the analysis of experimental results was also a basis in the 19th century for philosophical considerations concerning the mechanism of theory formation and the range of its validity. For example, John Herschel proved that natural laws and theories are formulated either by way of induction or by advancing a hypothesis, and William Whewell emphasized the role of prevailing views which, together with the observed facts, contribute to the formulation of natural laws and theories through induction. At the turn of the 19th century, Henri Poincaré claimed that apart from empirical components each scientific theory is bound to have some degree of convention [13].

Simultaneously, *positivistic empiriocriticism* was being developed. It rejected all abstract interpretations, and recommended the investigation of direct experience only. According to the Soviet historian of science O. V. Kuznetsova [37] it was a reaction to the dogmatism of many contemporary physicists whose views were often shaped by a literal interpretation of mechanistic models. Energetism, which developed on the basis of empiriocriticism, at the turn of the 19th century, denied the real existence of atoms and molecules (cf. Chapter 3).

In spite of enormous progress in the practical application of science in the last decades of the 19th century, scientists often felt discouraged. They seemed to believe that all the fundamental and unchangeable laws of nature had already been discovered and there was very little to be done in science except for some technical improvement of devices which could increase the accuracy of measurement or the production of new materials. Little attention was paid to a few still unsolved problems such as, for example, the arrangement of lines of atomic spectra, optical phenomena in vacuum tubes, the problem of radiation of black bodies, or the velocity of the Earth with respect to the cosmic ether.

1.6. The 20th Century

Solution of those neglected problems brought about a revolution in natural science. Experimental proofs of the complex structure of the atom were found at the turn of the 19th century. Albert Einstein proved theoretically the feasibility of changing the corpuscular form of matter into an energetic form. Some corpuscular properties (Planck's quanta and Einstein's photons) were attributed to the energetic form of matter, and wave properties (de Broglie's waves) were attributed to the corpuscular form of matter. Thanks to the development of statistical methods, the Democritean notion of accounting for macroscopic phenomena on the basis of microscopic phenomena subjected to indirect observation was being reborn.

New methods based on the theory of operators and matrices had to be devised for the mathematical description of those microscopic phenomena. This in turn led Erwin Schrödinger and Werner Heisenberg to the development of the methods of quantum mechanics; they involved new axiomatic elements and, at the same time, they represented a new version of the Pythagorean method of describing natural phenomena.

At the same time researchers came to a better realization of the fact that the laws and theories which they formulated on the basis of observation are valid only under the conditions under which such observations were made.

Relations obtained from deduction do not fully reflect the objective reality, but they may be merely regarded as successive approximations which describe it with an increasing degree of precision; new theories include an ever-increasing number of phenomena, and the former ones become their particular cases. Thus, what we call 'classical' laws turn out to be particular cases of more general laws which also control the behaviour of atoms and molecules.

The mechanism of the origination and development of scientific theories and concepts has been continually studied. According to the American philosopher Percy Williams Bridgman—a Nobel prize winner in physics—the formulation and verifiability of a scientific theory is largely restricted by the fact that we are unable to define all the parameters which accompany observations aimed at developing new theories. He also states that it is indispensable to adopt some non-analysable concepts, the number of which will change with time.

Modern scientists and philosophers also realize that our current cognition of the real world is not fully objective. The images by means of which we represent the surrounding world need not exist in nature in the same form as in our descriptions. The Austrian philosopher Karl Raimund Popper distinguishes three 'worlds': a world which exists materially irrespective of our cognition; a world of one's individual sensations, dependent on the amount of knowledge which one has accumulated throughout one's lifetime; and a world which is the outcome of the empirical, theoretical, and philosophical inquiries of mankind, i.e. a world of scientific concepts dependent on the current state of knowledge [13].

Similar views are also expressed by modern natural philosophers.

1.6. THE 20TH CENTURY

Hoimar von Ditfurth concludes that the genetic evolution of our brain has a decisive influence on the way in which we perceive external reality [38]. On the other hand, in a work published in Basle in 1935, the Polish bacteriologist Ludwik Fleck claims that the formulation of hypotheses depends in great measure on the cultural traditions of the milieu in which an investigator and thinker lives, and that the historical development of that milieu is reflected in the way the external world is conceived [39]. Another Polish scientist, Andrzej Fuliński, has recently supported this view:

(...) culture determines our reasoning, hence the way of creating and transmitting science [40].

From this standpoint it will be easier for us to understand the alchemists' approach to natural phenomena, which was quite different from ours; we can also find in the works of many modern investigators the echoes by bygone methods of reasoning.

The development of the analytical approach in particular disciplines has already gone so far that a reverse trend—a synthetic, holistic view of the world—is becoming more apparent. This process was initiated in physics in the previous century when its particular branches were successively merged into one; first mechanics and the science of heat gave rise to thermodynamics, subsequently thermodynamics was linked with electricity, magnetism, and optics.

The present development of scientific inquiry seems to favour a return to the holistic approach to nature which has been discarded since the time of alchemy. We are now beginning to realize that an analytic approach to particular phenomena and problems, which was quite successful over a long stage of the development of scientific inquiry, has now become inadequate.

Such an approach finds support in the recent advances in the natural sciences. The development of new research methods and an improvement in the accuracy of measurements have made it possible to observe phenomena brought about by particular fragments of molecules, as well as by single atoms or even their parts. We are not only able to discover the presence of such tiny parts of matter and their arrangement, but to study their interactions as well. This constitutes a basis for a uniform and holistic treatment of problems in the natural sciences, particularly in physics, chemistry, and biology. Of course, in this situation the traditional academic division into inorganic, organic and physical chemistry is no longer valid. Methods developed in those particular branches of chemistry are now being used to solve all the major problems of the whole of chemistry. We shall mention here, for example spectroscopic, chiral-optic, and dielectrometric methods, as well as those of irreversible thermodynamics, quantum methods of calculation *ab initio*, and methods of synthesis of large molecules.

The main current trends in chemical research include the following problems:

(1) the structure of condensed phase of crystalline bodies, alloys, and solutions;

(2) the structure of natural compounds and their synthesis;

(3) the mechanism of chemical reactions, including reactions that occur in living organisms;

(4) interphase phenomena;

(5) chemical sources of energy;

(6) synthesis of materials having desired properties;

(7) ecological issues.

As a rule, research is being carried out not by individual scientists but by teams of scientists. It is something of a paradox that, in the past, macroscopic bodies and molecular complexes were investigated by individual researchers and now large teams of investigators are engaged in research into individual molecules.

The American historian of science, Derek J. de Solla Price, has proved that the number of persons engaged in scientific research increases exponentially throughout the centuries, and if this rate were to continue the number of scientists at the turn of the 20th century would equal the number of the whole world population [41]. Therefore, there is a need to limit the range of research and concentrate efforts on selected issues because otherwise the number of scientists would be inadequate to carry out research in all possible directions.

A similar conclusion can be drawn from an analysis of the increase

1.6. THE 20TH CENTURY

in expenditure on research. In the past a scientist carried out investigation at his own cost or had a rich sponsor who enjoyed performing scientific experiments. Nowadays research costs exceed the financial abilities of individual persons. The cost of experimental installations is rising constantly, and advances in research techniques are so rapid that experimental facilities become outdated after a few years of use, i.e. the provision of up-to-date facilities enables one to obtain in the same period of time a greater amount of more exact data. It happens that the results obtained by the apparatus manufactured a few years ago are of less scientific validity than those obtained by more up-to-date apparatus. Due to its immense cost research must be financed by industrial companies, governmental, and—more often—international organizations, which nonetheless are unable to sponsor all scientific activity.

Thus science administrators, who lay out new directions of research, are becoming more and more important. Progress in science seems to depend on their decisions and recognition. However, science forms an integrated whole, advances in one field being determined by developments in a related one. Here lies the difficulty of science administrators; they have to predict which problems will be essential in the near future, which science should therefore be developed more intensively, and which sciences may be allowed to evolve at a slower rate. It is necessary to plan the development of science in its correct perspective. Research and intellectual potential should not be dissipated on the one hand but, on the other, scientific development must not be restricted to mere practicability.

CHAPTER 2

The Element

2.1. Antiquity

A natural phenomenon occurs when we discover that an object under observation has undergone a change. For instance, earth-like ore is transformed into reddish-brown copper. What is the observed object: ore or copper? What has undergone a change, and what has remained unchanged? The preserved historical evidence shows that such questions were asked by the Ionian philosophers as early as the year 600 BC. They observed a cycle of transformations: plants grow in earth, water, and air; then they are devoured by animals; and animals are resolved into earth, water, and air. However, what is the primary matter which has remained unchanged during these cyclic transformations

According to a legend retold by Aristotle and Plutarch, observations of the Nile flows which left behind a fertile soil induced Thales of Miletus to regard water as primary matter. During a shower rain of water falls from the air, hence air is rarefied water. And thin air is fire which rises into the air. Therefore, as a result of rarefication and condensation, earth, water, air, and fire could be transformed into one another. Their primary matter, however, is not changeable, but it manifests itself in different forms.

Other Greek philosophers sought a primary substance in different forms of matter. Anaximenes of Miletus considered rarefied air to be the primary matter—*arche* because a gentle breath of air is a sign of life. The Greek term *pneuma* used by Anaximenes was later translated into Latin as *spiritus*, which means air, breath, and also spirit or soul, which was held to give life, and was later opposed to body.

On the other hand, some Greek philosophers treated air as vacuum. Anaxogoras and Empedocles proved, however, that air is different from vacuum and vapour because in water clocks water enters the

2.1. ANTIQUITY

space occupied by air which escapes from the clock. In turn, Heraclitus, who emphasized eternal changeability, recognized fire as primary matter because it continually changes its appearance.

Empedocles synthesized the contemporary views; he held that there exist four primary substances: earth, water, air, and fire. In accordance with a view that the nature of animate beings is analogous to that of inanimate beings, Empedocles also distinguished in the human body for elementary fluids or 'humours' which corresponded to the four primary substances. Thus fire corresponded to red blood (Greek *sanguis*), air to yellow bile (*chole*), water to white phlegm, and earth to black bile (*melaina chole*). The four humours of Empedocles have been preserved up to the present day in psychology in order to characterize human temperaments; we speak of sanguine, choleric, phlegmatic, and melancholic dispositions.

Thus the Ionian philosophers thought that there exists one primary matter whose form is continually changing; on the other hand, as we have seen in Chapter 1, the Eleatics emphasized the unchangeability of primary matter. We shall discuss its texture in more detail in the next chapter; now we shall only say that as early as the 6th and 5th centuries BC the problem of the divisibility of matter was raised. Democritus set forth a hypothesis that primary matter occurs in the form of infinitely diversified, indivisible, minute particles called *atoms* between which is the void. Democritus attributed the multiplicity of the forms of matter to the union and separation of atoms. In fact, this view explained what had already been known, but no practical conclusions could be drawn from it. Therefore, for a long time the atomist conception appeared to be less fruitful than those based on Aristotle's considerations, which will be discussed later.

Let us note that primary matter (*arche*) was a concept which in fact did not denote anything concrete; however all observable things were believed to be composed of it. On this assumption Plato based his objective idealism, referred to in Chapter 1. Plato exerted an enormous influence on his pupil Aristotle, the most outstanding philosopher of antiquity. The works of Aristotle are the main source for the study of the whole of Greek science between the 6th and 4th centuries BC. He collected all scientific theories existing in his lifetime and presented them critically from his point of view. For the present-day historians of physics and chemistry, the most important works include: 'Physics', 'On Generation and Corruption', 'On the Heavens' and 'Meteorology'. Following the thoughts contained in these works, we should bear in mind that they reflect a gradual development of Aristotle's views and do not provide a consistent exposition of one philosophical system.

Aristotle begins 'Physics' with the following consideration:

When the objects of an inquiry, in any department, have principles (*archai*), conditions (*aitia*), or elements (*stocheia*), it is through acquaintance with these that knowledge, that is to say scientific knowledge, is attained. For we do not think that we know a thing until we are acquainted with its primary conditions or first principles, and have carried our analysis as far as its simplest elements. Plainly, therefore, in the science of Nature, as in other branches of study, our first task will be to try to determine what relates to its principles [23, 184a].

It follows from the above quotation that Aristotle regarded the elements as real sources of knowledge, in contrast to Plato for whom reality was not made up of things but of ideas.

Let us remember that according to what Aristotle wrote in 'Metaphysics' the three sources of knowledge and comprehension—principles, conditions, and elements—denote slightly different concepts, although their meanings overlap strongly. They refer not only to components but also to their properties and transmutations.

However, Aristotle held that the primary matter (*yle*), the substratum, which is the basis of all other substances, is not accessible to our senses; we perceive only a moulded matter. On the other hand, the form of matter is determined by contrarieties, i.e. pairs of contrary properties. It is in accordance with a view cited in Chapter 1 that the existence of contrarieties is the determinant of all phenomena occurring in Nature.

In consequence, Aristotle rejects a view which resulted from unitarianism that there exists only one primary substance "for there cannot be one contrary" [23, 189a]. He also rejects the view advocated by Democritus that every substance consists of identical atoms specific only for this particular substance because multiplicity of primary matter cannot "be innumerable because, if so, being will not be know-

2.1. ANTIQUITY

Fig. 4. Symbolic representation of Aristotle's four elements, from a work of Peter Bonus *Margita preciosa novella*, 1546, acc. to [9].

able" [23, 189a]. Basing himself on dualistic conceptions, Aristotle concludes that there must exist two principles which create contrarieties, and also their substratum which "is different from the contraries for it is itself not a contrary" [23, 190b]. In 'Physics' Aristotle seems to link both the unitarian and dualist conceptions and comes to a conclusion that recognition of the existence of three elements will be most justified.

The most comprehensive discussion of the elements is given by Aristotle in his next work entitled 'On Generation and Corruption'. At the outset he recalls the views of earlier philosophers according to whom there exists only one primary matter—earth, water, air, or fire; and also the opinion that these four substances are four different forms of one primary matter. Next he states that "the primary materials, whose change (whether it be association and dissociation or a process of another kind) results in coming-to-be and passing-away, are rightly described as originative sources, i.e. elements" [23, 329a]. However, such primary material always exists together with contrarieties and therefore it is "the origin of the so-called elements". On this basis Aristotle distinguishes three generative sources which account for the properties of elements: "firstly that which is a potentially perceptible body, secondly the contrarieties (I mean, for instance, heat and cold), and thirdly fire, water, and the like" [23, 329a].

Thus, we can see that according to Aristotle being does not manifest any particular properties but it can take them on. These properties constitute contrarieties which form being into the particular elements. Hence his speculations about the contrarieties are very important. Aristotle lists many pairs of contrarieties: hot–cold, dry–moist, heavy–light, hard–soft, viscous–brittle, rough–smooth, coarse–fine. Of these he distinguishes the two main pairs: hot–cold, and dry–moist. The Greek term for 'moist'—*hydron*—also implies 'fluid'. Next Aristotle goes on to prove that the remaining contrarieties can be reduced to these two basic pairs because, for example, a fine body fills up a vessel in an analogous manner as a moist or liquid body. Of the two main pairs of contrarieties he recognizes the pair 'hot–cold' as active, i.e. having power to act, because "hot is that which 'associates' things of the same kind (for 'dissociating', which people attribute to Fire as its function, is 'associating' things of the same class, since its effect is to eliminate what is foreign), while 'cold' is that which brings together, i.e. 'associates', homogeneous and heterogeneous things alike" [23, 329b].

On the other hand, the pair 'dry–moist' implies 'susceptibility', hence it is passive. Finally, Aristotle concludes:

> It is clear, then, that all the other differences reduce to the first four, but that these admit of no further reduction. For the hot is not essentially moist or dry, nor the moist essentially hot or cold: nor are the cold and the dry derivative forms, either of one another or of the hot and the moist. Hence these must be four [23, 330a].

Similar considerations are also found in 'Meteorology' [23, 339a].

As can be seen, Aristotle lays much emphasis on the distinction of elementary qualities—properties: two pairs of contrarieties: hot–cold and dry–moist. On this basis he develops his conception of the science of the elements in his treatise entitled 'On Generation and Corruption':

> The elementary qualities are four, and any four terms can be combined in six couples. Contraries, however, refuse to be coupled: for it is impossible for the same thing to be hot and cold, or moist and dry. Hence it is evident that the 'couplings' of the elementary qualities will be four: hot with dry and moist with hot, and again cold with dry and cold with moist. And these four couples have attached themselves to the apparently 'simple' bodies (Fire, Air, Water and Earth) in a manner consonant with theory. For Fire is hot and dry, whereas Air is hot and moist (Air being a sort of aqueous vapour); and Water is cold and moist, while Earth is cold and dry [23, 330a, b].

Thus, according to Aristotle, every element carries two basic qualities, one of which is active and the other passive. No element, however, can carry two qualities of the same kind. Hence two pairs of contrary qualities generate the existence of only four basic elementary substances.

However, Aristotle was not consistent in his presentation of the role of the elements in the texture of matter. In his next treatises we find contrary statements. Thus in the treatise 'On the Heavens', he writes:

> An element, we take it, is a body into which other bodies may be analysed, present in them potentially or in actuality (...) and not itself divisible into bodies different in form [23, 302a];

an in the treatise 'On Generation and Corruption', we find a statement:

> In fact, however, fire and air, and each of the bodies we have mentioned, are not simple, but blended. The 'simple' bodies are indeed similar in nature to them, but not identical with them [23, 330b].

Every element carries two qualities, one of which is dominant. Thus, fire, which is an 'excess of heat' bears the quality of 'hotness'; moisture is attributed to air, and dryness to earth.

For Aristotle earth means all solid bodies; water—all liquids; and air—all substances that are today called gases or vapours, including hot steam, which upon cooling forms water. Each of these three elements, as a matter of fact, represents one of the three states of aggregation. However, fire has a particular significance; it is not only one of the elements, but, moreover, it is an agent that can excite activity; an agent that, as we have already mentioned, blends homologous bodies, and removes foreign bodies. We shall discuss the role of fire in great detail in Chapter 5.

Further considerations of Aristotle contained in the treatise 'On Generation and Corruption' confirm the conception of elements as the representatives of the states of matter: "(...) the coming-to-be of the simple bodies is reciprocal (...)", and "all of them are by nature such as to change into one another" [23, 331a]. Thus, Aristotle argues that air can change into both water and fire. We shall deal with the problem of such transformations in Chapter 5 when discussing the ability of substances to react.

In 'Meteorology' Aristotle attributes one more property to these four element-bodies—vertical motion upwards and downwards:

> We have already laid down that (...) these four bodies owe their existence to the four principles, the motion of these latter bodies being of two kinds: either from the centre or to the centre. These four bodies are fire, air, water, earth. Fire occupies the highest place among them all, earth the lowest, and two elements correspond to these in their relation to one another, air being nearest to fire, water to earth [23, 339a].

Thus, all bodies occupy a natural position with respect to the Earth, and when out of balance, they move linearly upwards or downwards in order that they may return to their natural position. This kind of reasoning prompted Aristotle to introduce one more, i.e. the fifth element. Besides vertical motion the ancients knew of the rotary motion of the stars and planets, which never fell upon the Earth. They were believed to orbit around the Earth above the sphere attributed to fire. It would have been easy to assume that they move in a vacuum, but Aristotle rejected the concept of vacuum as a homogeneous area devoid of any matter. Thus the heavenly bodies move, according to Aristotle, in some materials medium. In the treatise 'On the Heavens', he states:

(...) if the heavenly bodies moved in a generally diffused mass of air or fire, as everyone supposes, their motion would necessarily cause a noise of tremendous strength and such a noise would necessarily reach and shatter us [23, 291a].

It follows clearly from the above considerations concerning rotary motion in a material medium that, except for the four substances existing on the earth, there exists a bodily substance which is more divine and primary than all the others. In order to emphasize that the primary body is different from earth, fire, air, and water, the ancients called it 'ether' (*aither*), "implying that the primary body is something else beyond earth, fire and water, they gave the highest place a name of its own, either, derived from the fact that it 'runs always' (*aei dhein*) for an eternity of time" [23, 270b]. Aristotle pointed out that the very term 'ether' had existed earlier; it referred to Zeus—the spirit of the world; and according to Empedocles, it referred to an unspecified, most subtle matter. The Pythagoreans also regarded ether as the fifth element.

Thus, to the four 'earthly' elements, Aristotle added the fifth—heavenly ether, which was called by the Romans *quinta essentia*. This term gave rise to the present-day word 'quint-essence'.

As we have already said, Aristotle's views do not form a uniform whole. For example, in 'Physics' he states that only three elements should exist; in the treatise 'On Generation and Corruption', he accounts for four elements; and in the treatise 'On the Heavens', he introduces the fifth element—ether. He regards the element as an indivisible body, but elsewhere he states that air can include both hot and cold steam (since it is derived from water) as well as hot and dry smoke.

All statements of Aristotle were repeatedly commented on and interpreted in the Middle Ages; they exerted an enormous influence on the development of thought at that time. Traces of those views have been preserved up to the present day in many beliefs and opinions.

2. THE ELEMENT

Our considerations concerning Aristotle's texts will be supplemented by an analysis of the origin of the term 'element'. Aristotle used the word *stocheion* which meant for him both element and principle. In Greek this word meant primarily a speech sound, a letter, but it had wider connotations, too. The Greek *stocheion* corresponds to the Latin *elementum* containing three consecutive letters of the Latin alphabet—*L*, *M*, *N*, which, at one time, were the first letters of the Latin alphabet. The word 'element' was subsequently adopted by almost all the European languages.

Thus, we can see that throughout the whole period of Greek science all the earthly bodies were believed to consist of four elements: earth, water, air, and fire. In proof of that view the Greeks argued that when a tree trunk is burning we can see a flame (fire), smoke (air), juices (water) dripping off the trunk, and after burning out, uninflammable ash (earth) is left [32, p. 24]. Of course, these four substances could not have been regarded as pure elements because, according to Aristotle's view, the concept of the elements concerned non-formed matter, whereas we observe formed matter, e.g. in the form which appears to us when we watch a burning trunk. Each of the above-mentioned elements was believed to dominate respectively in each of the observed substances.

The four elements of unshaped matter were believed to be—as we have seen—carriers of elementary qualities: heat, dryness, cold, and moisture. Addition of any of the elements to a particular body would have changed the relative amounts of the particular elements contained in that body, and hence its properties would have been changed. For example, when dry and hot fire was added to water (whose main component was the cold and moist element—water) hot and moist air (steam) as well as cold and dry earth (precipitated salts) were obtained. This was due to the fact that common undistilled river or spring water was vaporized.

2.2. Alchemy

The view that by changing the proportion of the elements one can change the properties of bodies was the basis of the attempts aimed at ennoblement of metals and minerals. On such reasonings were based the efforts of the Arab and European alchemists to transform base metals into gold. We have seen that in the early Middle Ages everything that looked like gold was actually believed to be gold. Hence, it was believed that bronze is more akin to gold than pure copper, and therefore attempts were made to ennoble bronze in order to obtain gold.

The Arab alchemists did not in fact endeavour to transform all base material into noble material, but they only attempted to give it a superficial appearance or—in Aristotle's terminology—a form of noble material. Therefore, ways were sought to make imitations of precious stones, and to dye fabrics more cheaply. These efforts contributed to the significant growth of the chemical art, as has been mentioned in Chapter 1.

The Arab alchemist Geber (9th century) is credited with saying that mercury is a metal despite its liquidity and volatility. He is believed to have noticed that sulphur and gold have similar colours. Hence, an opinion was spread that one could obtain a gold-like metal by mixing metallic mercury with yellow sulphur. This played a significant role in European alchemy.

European alchemists in the late Middle Ages resumed the Arab attempts to transform bodies. They, however, failed to realize that the Arabs wanted merely to change the external appearance of bodies. They believed in fact that they would succeed in transforming base metals into gold.

In the first centuries of the modern era, nine substances which are now called elements were known, but they were not then regarded as simple bodies. They included seven metals: gold, silver, mercury, copper, iron, tin, and lead (zinc was known only in alloys), and sulphur and carbon besides, the latter in the form of diamond, which had the strange property of vanishing. On the other hand, the four elements of Empedocles and Aristotle were believed to be the elements, bases, and essences (these terms being used interchangeably) of all substances. It was not stated explicitly that they were bases because the classification of a substance was made, as we have mentioned, by its external appearance. Yet the Arab alchemists had discovered that particular earths of similar appearance could have different properties, but the base-earth was a substratum which, of course, could assume various forms. Different properties of real earths did not therefore contradict the views founded on Aristotle's theories.

To the four Aristotelian elements, carriers of contrarieties, the Arab alchemists in the 10th century added two more substances: mercury, a carrier of metallicity and inflammable sulphur, which was recognized as a carrier of combustibility. Each of these new elements, together with salt, a carrier of solubility, recognized by Paracelsus in the 16th century as an element, therefore, represented only one quality which did not have a contrariety. The dualist conception that all metals contain two of the 'new' elements—mercury and sulphur—was adopted by the European alchemists, and in the 13th century it became widespread thanks to two fathers of the Church: Albert of Bollstädt, known as Albertus Magnus, and Thomas Aquinas. We shall present these views according to *Ein wahrhaftige Lehr der Philosophie*, a Philosophical Treatise probably written in 1423 by a certain German philosopher:

(...) in Nature matter is primary being concentrated of the four elements only according to the recognition and qualities of Nature, and this matter being referred to by philosophers as mercurius or quicksilver; but notwithstanding that it is imperfect due to an excess of sulphuric earth which is too filmy and combustible, and also due to excessive moisture which is all concentrated from the four elements under the influence of the supreme planets (...). Since Nature always strives towards the supreme, and to an end, which is designed by the Creator of all things, it will not allow matter to remain unfinished, and by the continuous mixing of the four qualities of the four elements Nature will fulfil its task and through the falling heat of the Sun and natural warmth the said matter shall be moved together with its inner sulphurousness; thus it rises in the cracks and veins of the earth in the form of vapours and smokes. And since such a vapour or smoke cannot escape and is confined, it must penetrate into certain slimy earthy greasiness and impure sulphurousness which can be found in the earth veins; and furthermore, then this matter absorbs its liquidity and foreign impurities, so it becomes thereby more impure, and this is the cause that it assumes different colours before it reaches its proper purity and colour [42].

Thus, the variety of metals was a result of the contamination of the pure elements: mercury and sulphur. On the other hand, it was the

2.2. ALCHEMY

wandering stars, i.e. the Sun and the Moon, and the planets, which determined the generation of a particular metal at a given moment. In Chapter 1, we have seen that a combination of contrarieties was, according to the alchemists, a prerequisite for the formation of a new entity. Hence, the formation of metals must have been an effect of such a cause. This opinion found confirmation in another statement of the German philosopher:

Mercury is the mother of all metals due to its cold and moisture (...), whereas sulphur is the father of all metals due to its heat and dryness.

Furthermore, the author explains why ores of metals are most frequently found in the mountains. He believes that God created the Earth "flat, vile, and fat, without gravel, sand, or stone, without mountains and valleys", but under the influence of the planets stones appeared on the Earth's surface and ores in its interior. Under the influence of the Sun's heat a strong smoke and misty vapour carried upwards parts of the 'Earthy estate' and formed hills and mountains where "earth was favourably saturated, transformed, and mixed with heat, cold, and moisture; the best ores are found there".

This is illustrated by the title page of the treatise (Fig. 5). From the depths of the Earth vapours are rising which contain, as the alchemical signs suggest, mercury and sulphur. The vapours are mixed in the interior of a mountain and thus form an ore which is extracted and processed by man. The whole process is explained by a philosopher whose figure, with a book in his hand, dominates the right side of the illustration.

Formulations almost identical to those quoted above can be found in Chapters 3 and 4 of 'A treatise about the philosopher's stone' by Sendigovius, a work also referred to as *Novum Lumen Chymicum*, published for the first time in 1604:

The primary matter of metals is twofold, but neither can produce metal without the other (the former being mercury, the latter sulphur—R.M.) (...) all bodies come from liquid air or vapour the elements of which ooze deep into the Earth through constant motion. Next Nature's Archeus received it all and sublimes it through the pores of the Earth, and by virtue of his power he endows it in every place, thanks to which due to diversity of places different things are born. There are some who believe that Saturn has a different seed, and gold has a

Fig. 5. Formation of ores from the vapours of mercury and sulfur under the top of a mountain; the title page of one part of the work *Dyas Chimica Tripartita*, 1625.

2.2. ALCHEMY

different seed too, and other metals likewise. This conjecture is fallacious for there exists only one seed, this same is in Saturn and gold, this same is in silver and in Mars, etc. However, the place on the Earth was different, if you understand me well, and on this account Nature ceased to act sooner in the Moon than in the Sun, and in the other metals likewise, (...) the purer is the place the more beautiful are the metals that are found [30, pp. 164, 166].

Although Sendigovius refers to silver as the Moon, to gold as the Sun, and lead as Saturn, the diversity of metals is in his opinion caused by the place where "this vapour, called philosopher's Mercurius, is adjusted and mixed with greasiness and then carried up to the surface".

In connection with the two alchemical texts cited above, let us direct our attention to two problems: (1) the relation of metals to planets; (2) the relation of the four Aristotelian or peripatetic elements to the alchemic (spagyric) elements. This latter problem will be discussed in detail a little further on in this book. Now we shall deal with the seven metals and the corresponding wandering stars.

Attribution of particular metals to the wandering stars dates back to the times preceding the period of Greek science. It was the Babylonians who, in some case, had already attributed alloys and not pure metals to stars. The Greeks replaced *electron*—an alloy of gold and silver—by mercury, and bronze or an alloy of copper and tin—*speculum*—by pure copper. We have already mentioned that metals were originally found in meteorites which fell from the sky, and their

Table 1. Seven alchemical combinations*

Metal	Planet	Archangel	Part of body	Flower**
lead	Saturn	Orphiel	spleen	marjoram
tin	Jupiter	Zachariel	liver	lily
iron	Mars	Samuel	bile	lotus
gold	Sun	Michael	heart	buttercup
copper	Venus	Anael	kidneys	narcissus
mercury	Mercury	Raphael	lungs	white violet
silver	Moon	Gabriel	brain	rose

* According to the diagram *Mundus elementaris* in: *Dyas Chimica Tripartita*, Luca Jennis, Franckfurt am Mayn 1625 (cf. Fig. 6).
** According to Berthelot M., *Introduction a l'étude de chimie*, Paris 1938.

cult, widespread among primitive peoples, favoured associating them with gods. Analogously, particular planets were also associated with gods.

Thus seven metals were attributed to seven wandering stars and seven gods as well as seven parts of the human body. The Christian alchemists replaced pagan gods by archangels. In his study of alchemy, Berthelot showed that alchemists had also distinguished seven flowers and seven odours, each of them being attributed to a respective metal and planet. This combination is shown in Table 1.

It should be remembered that in the Middle Ages a particular mystical significance was attributed to the number seven. Even the inquisition reproached Galileo in 1616 for claiming to have discovered new wandering stars (i.e. Jupiter's moons) since, it was held, not more than seven wandering stars existed.

It seems worthwhile to analyse the above tabulation of metals and planets (cf. Fig. 6). Attribution of the two shiniest metals and most resistant to blackening—gold and silver—to the two largest wandering stars—The Sun and the Moon—seems quite natural. It is also quite obvious that iron is associated with the planet devoted to the god of war—Mars; *speculum*—mirror alloy or copper used for the manufactures of women's ornaments was logically associated with the planet named after the goddess of love—Venus; the origin of the remaining two and least noble metals—lead and tin—was attributed to the least magnificient wandering stars—Jupiter and Saturn. The signs which symbolized the planets were also used to symbolize the respective metals. Until today only two of those signs have remained in science: ($♂$) the sign of Mars—iron denotes male and ($♀$) the sign of Venus denotes female in biology.

Such an attribution of metals to planets was commonly recognized by alchemists and was not questioned for centuries. The sequence of the planets was known and it imposed a definite sequence of metals. Attempts to justify such an arrangement of metals by their chemical or electrochemical properties, as some writers suggest now, do not seem to be quite convincing. On the basis of the sequence of planets–metals, it is difficult to suspect that the Greek philosophers knew of the electromotive series. It is clear that the noble metals are placed at the

2.2. ALCHEMY

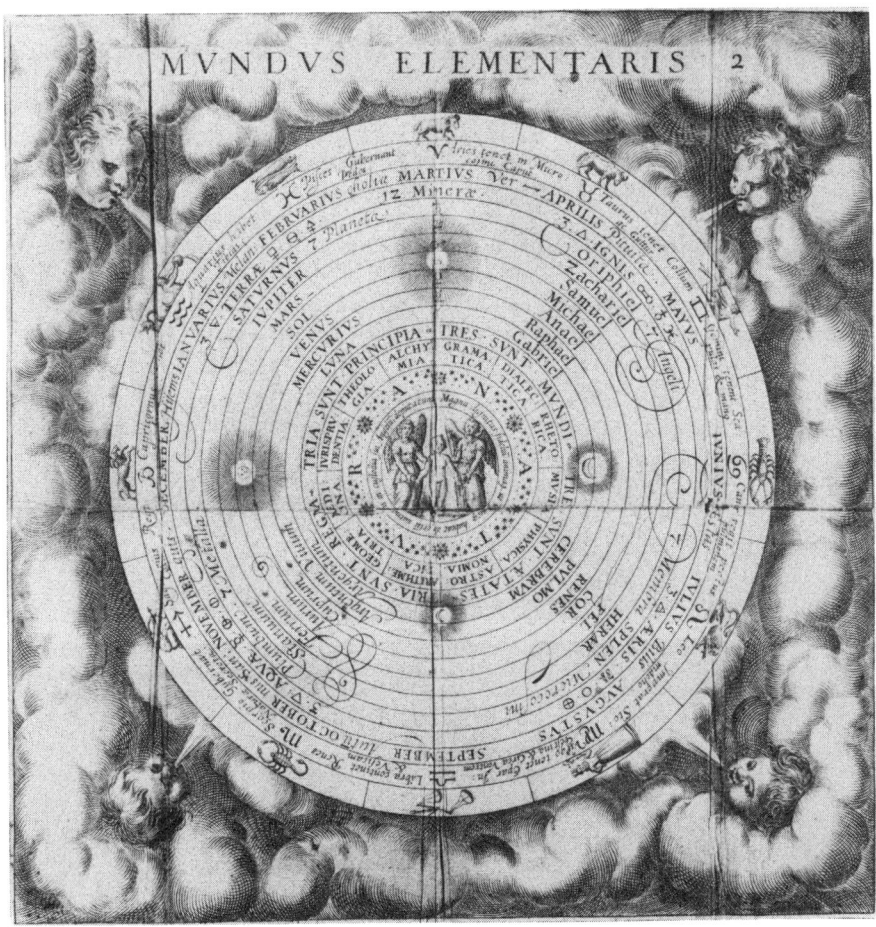

Fig. 6. Elementary world, from *Dyas Chimica Tripartita*, 1625.

bottom of this series, and they are thus attributed to the brightest wandering stars.

The chemical transformations dealt with by alchemists in the 13th, 14th and 15th centuries concerned mainly mineral bodies. In the first half of the 16th century, the Basle alchemist Paracelsus extended the spagyric conceptions to the kingdom of vegetables, animals, and man himself, and he treated diseases on the basis of his conclusions. As has

44 2. THE ELEMENT

been mentioned in Chapter 1, iatrochemistry came into being in this way. We shall illustrate Paracelsus' views by a few quotations from his works:

Separation of things which grow from the earth and are inflammable, such as fruit, herbs, flowers, leaves, grass, roots, trees, etc., can be effected in a variety of ways. By distillation phlegm is first given off, then, mercury, next oil, fourthly sulphur, and their salt is left. When all these separations are made in accordance with the spagyric art, we obtain perfect and magnificient medicine for both internal and external use (...).

As to the way God created this world, we may assume the following relation. At first he reduced it to one body while the elements developed. This body is

PHILIPPUS
THEOPHRASTUS
BOMBAST

Fig. 7. Paracelsus, acc. to [7].

2.2. ALCHEMY

made up of three components: Mercury, Sulphur, and Salt, so that three compose one body. All bodies which are in four elements are composed of these three components [43].

Next Paracelsus analyses in what way the element–air consists of three spagyric elements and how smoke, i.e. air, is made in the process of burning various bodies and those three spagyric elements contained in them. In his opinion salt plays a significant role in animate organisms. He devoted to this problem a treatise entitled 'On Salt and Substances Conceived of as Salt'. He writes:

God has raised man so high and led him to such limits of necessities and necessaries that he cannot live without salt which is indispensable in his nourishment and fare (...).
Man consists of three things: sulphur, mercury, and salt (...). Further, since man is made up of parts, he is subjected to decay, and he cannot avoid it unless God provides him with an innate balm which is in three things. Thus, salt protects man from decay; the part where salt is scarce will decay (...). And since all the things created, all substances consist of these three, it is necessary that they are maintained and conversed by nutrients contingent upon their kind (...). Therefore, it becomes evident by analogy that man himself must be nourished in this same way, i.e. sulphur must be fed with sulphur, the nutrient of mercury is mercury, and innate salt is the nutrient of salt, and the three maintain and preserve man in his kind. Whatever burns is sulphur, whatever is wet must be mercury, and what is the balm of these two will be salt [44].

It was the considerations of Paracelsus about reinforcement of the particular constituents of the human body by nutrients containing sulphur, mercury, or salt, respectively, that became a basis for the application of mineral compounds in pharmacy.

Mercury was considered to be a metal and it was also recognized as a carrier of metallicity. Moreover, it was liquid at room temperature, and when heated it evaporated. Therefore, in some writings we can find mentions of mercury as a symbol of liquidity or volatility. However, Aristotelian water possessed the same properties. On the other hand, it was recognized that combustibility was a characteristic property of sulphur and that is why the alchemists considered sulphur to be a carrier of combustibility. Yet numerous substances were known to be neither volatile nor combustible; many of them would dissolve in water. They were mainly minerals, also generally referred to as 'salts'. Hence Paracelsus treated salt as a separate element which was to be a carrier of non-inflammatory or solubility. Nonetheless, those non-inflammable minerals were regarded by the peripatetics as earth.

Let us note that these earths/salts were relatively passive solid bodies whose properties differed significantly from caustic solid bodies —alkalis—then obtained by the evaporation of lyes from the ashes of various plants. Thus, such lyes were included in the kingdom of plants. As late as about 100 years later alkalis obtained by leaching were regarded as bases of salt, and salts as lye bases neutralized by acids. Acids were considered to be volatile substances given off, as we have already said, during the roasting of salt. However, even then the name 'salt' referred to a general notion because, as late as 1668, Otto Tachenius asserts:

(...) all salts which annihilate acids are called by one general term, alkalis, which are to be found not only in the kingdom of plants but in the kingdoms of animals and minerals as well (...) [45].

Although it was realized that some substances regarded as salts were obtained by treating acids with alkalis, salt was still considered by some chemists in the 17th and even 18th centuries as an element. However, there was a certain difference between the peripatetic and the spagyric elements; the former being carriers of two opposing qualities, whereas the latter were carriers of only one quality. Besides, oddly enough, the three spagyric elements were believed to consist of four peripatetic elements. In some alchemic writings sulphur was even identified with fire, mercury with water, and salt with earth.

In a textbook *Alchemia*, published in 1597, the alchemist and Paracelsist Libavius attempted to reconcile the contradicting theses about the elements by stating that, depending on the kind of approach, one can obtain different bases for the classification of simple bodies. Thus, it can be assumed that only one element exists—primary matter (Greek *yle*); when considering the properties of metals, it will suffice to assume the existence of only two elements: mercury and sulphur; for other bodies three elements are essential: salt, mercury, and sulphur; whereas, when considering the properties of earthy bodies in

2.2. THE ALCHEMY

general, one can take into account the existence of four elements: earth, water, air and fire, or—possibly—the number of elements may be extended to five by adding ether after fire. Libavius went on to argue that, considering different classes of the division of the elements, then the particular elements within each class could be arranged according to their rising 'virtue'. It should be borne in mind that the Latin term *virtus* employed by Libavius has a very wide meaning. It denotes all positive properties with reference to both the character of man and to the qualities of inanimate objects. Thus, in the opinion of that alchemist the greater the 'virtue', i.e. ability or reactivity of the elements, the lesser was their 'materiality', i.e. passivity [31].

As we have already mentioned, in the period of alchemy ideas were represented by symbols, and so were the elements. Such symbols, together with maxims, were attributed to particular philosophers, some of them were mythical or obscure figures, and in the form of emblems were published in various books. Most of them were collected by Johann Daniel Mylius. In the work entitled *Hermetico-Spagyrisches Lustgärtlein* published in 1625, as many as 160 emblems were contained. In the book credit is given to Heraclitus for his maxim: "Fire is the origin of all things", to Geber ("an Arabian philosopher") for his opinion that "the whole of Nature is based on the sun and salt", and to Thomas Aquinas ("an Italian chymist") that "Both Nature and Arts create metals from sulphur and mercury" [46].

The above quotation expresses opinions which are vague and contradictory at times. It points to a great confusion which prevailed in chemistry at the turn of the 16th century. It was due to the fact that these opinions were founded on the authority of fallacious assumptions made by the Greek philosophers who had formulated them from a limited number of observations. Two thousand years later the number of observations had increased dramatically and it was no longer possible to deal with the results obtained within the old conceptual framework. At the turn of the 16th century investigators already realized the distinction between different salts/earths, different kinds of water/air, and they even assumed different forms of fire; this was graphically denoted by certain modifications of the main alchemical signs. Different earths meant different minerals; different waters meant different liquids; different airs were vapours, and different fires meant gases given off under the effect of fire from solid bodies.

2.3. Weight Methods and the Concept of the Element

The development and spread of quantitative methods at the turn of the 16th century resulted in a significant change in the methods of reasoning. The belief in the transmutation and ennoblement of metals became weaker. Alexander Zuchta (von Suchten) of Gdańsk proved quantitatively that in many chemical processes gold is not formed but given off in the amounts previously introduced. Daniel Sennert and Jan Baptista van Helmont found that copper is precipitated on an iron bar immersed in the solution of copperas (copper sulphate) not as a result of the transformation of iron into copper, but due to the copper which was already in the solution. Van Helmont also held mercury vapours to be the same mercury as in the liquid state, and silver after dissolving in *aqua fortis* (nitric acid) still remained silver.

The simplest quantitative method used in chemistry was a weight method, which was particularly applicable to alloy analysis. Cracow was one of the most advanced assay centres in the 15th century. In the 16th century, on the basis of observed changes in the weight of metals subjected to heating, attempts were made to explain the mechanism of the processes involved. Fire which accompanies burning had been known, as we have mentioned, since prehistoric times. It was realized that the ash which remains after a plant substance has been burnt weighed less than the substance being burned. Since the ancient times vegetable substances had been known to undergo another process which is a milder source of heat than combustion. This was fermentation, whose name according to Georg Ernst Stahl, the creator of the phlogiston theory [47, p. 9], is an abbreviation of the name *fervimentatio* (Latin—*fervo*—to heat). As effervescence of a substance is the property of fermentation, i.e. liberation of 'air' (gases), the product of fermentation must have been lighter than the substances undergoing this process, although to a much lesser degree than in the case of the same substance being burned.

2.3. CONCEPT OF THE ELEMENT

On the other hand, heated metals, or even metals left in air, were soon covered with a deposit called calx, and the process was called calcination.

As early as the 5th century Olympiodoros knew that this calx weighed more than the metal of which it was made. In 1560, Gerolamo Cardano asserted that lead having been transformed into white or red powder weighs 1/13 more "due to the blue fire given off, just like animals become heavier after death due to the removal of the soul" [16, vol. II, p. 13].

The 15th and 16th centuries marked the development of boiler making and smithery. During heat treatment the formation of scale was a considerable impediment in the work of craftsmen. This problem was investigated by the French chemist Jean Rey, who, in 1630, published a description of his experiments concerning an increase in the weight of tin and lead during heating. Like Galileo, Rey criticizes Aristotle's view that the upward movement is the natural movement of air and fire. He states:

Air and fire have their own weight. They are ponderable (...). This consideration has led me to the discovery of weight in all the elements and this consideration will allow me to negate a presumption that the elements being transformed into one another lose or gain weight.

He provides the following argumentation: "If we take the lightest portion of earth and change it in a natural way into water, then this smallest portion of water will not have a smaller weight". Next he goes on to say: "Weight is closely linked with the primary matter of the elements which cannot be devoid of it" (i.e. weight). The elements which Rey has in mind are peripatetic; they can change into one another. Like many later scientists (even in the 19th century), Rey did not distinguish the concepts of mass, weight, and density: this should be remembered when analysing his writings.

However Rey was convinced that fire could condense homogeneous bodies, because after distillation of wine or turpentine he found a constant residue in the flask. Therefore, he explains the observed increase in the weight of calcinated tin and lead in the following way:

(...) this increase in weight comes from the air, which in the vessel has been rendered denser, heavier, and in some measure adhesive, by the vehement and long-continued heat of the furnace: which air mixes with the calx (frequent agitation aiding) and becomes attached to its most minute particles: nothing other than water makes heavier the sand which you throw into it and agitate, by moistening it and adhering to the smallest of its grains [48].

In the cited text Rey employs the Aristotelian conception of the elements and refers to metal as earth. Moreover, like many later investigators, he fails to draw a distinction between a chemical process and physical mixing. Many science historians regard Rey as a forerunner of Lavoisier's theory of combustion, but let us note that, according to Rey, it is not metal but air that becomes heavier in the process of heating and then it is 'heavy air' that mixes or combines with metal, which is observed as an increase in the weight.

An analogous result of a similar experiment was explained by Boyle by the fact that the 'ponderable' particles of fire were attached by a metal as they penetrated through the walls of the vessel. But Boyle performed his experiment in an erroneous manner. In fact he calcinated the metal in a sealed flask, but he weighed it after opening the flask, and he failed to notice a hiss of air forcing into the flask. This mistake was pointed out in 1732 by the Dutch chemist Herman Boerhaave, who heated mercury in a sealed vessel for fifteen years, and he did not notice any change in its weight, and in 1744 by the Russian scientist Mikhail Lomonosov who thoroughly studied the works of the Dutch investigator. Both scientists held that fire has no weight (i.e. it is imponderable).

The above experiments show that quantitative methods began to play a dominant role in chemistry. Van Helmont also tried to approach the problem of Aristotelian elements in a quantitative way. In his work, published in 1648, he attempts to prove that water is the primary matter. He describes the following experiment:

I took an Earthen Vessel, in which I put 200 pounds of Earth that had been dried in a Furnace, which I moystened with Rain-water, and I implanted therein the Trunk or Stem of a Willow Tree, weighing five pounds: and at length, five years being finished, the Tree sprung from thence, did weigh 169 pounds, and about three ounces: But I moystened the Earthen Vessel with Rain-water or distilled water (alwayes when there was need) (...). I computed not the weight of the leaves that fell off in the four Autumnes. At length, I again dried the Earth of the Vessel, and there were found the same 200 pounds,

2.3. CONCEPT OF THE ELEMENT

wanting about two ounces. Therefore 164 pounds of Wood, Barks, and Roots, arose out of water only [49].

However, van Helmont's quantitative experiments could not save the outdated conception of the element. Joachim Jungius (Jung), whose fundamental work *Diocopiaephysicae minores* was published in Germany in 1642, considers which bodies are really simple. Although he does not provide a clear answer, he writes:

By a proper skill metals can be made to appear like common earth or sand, and yet again a simple metal can be recovered by melting. Is earth therefore a pure body? [50].

Re vera similaria, i.e. 'bodies which cannot be more decomposed in bodies of different properties under the effect of fire', include, according to Jungius, water, rock-salt, *sal urinae* (today NH_4OH), gold, silver, mercury, sulphur, and talc (magnesium silicate). Jungius rejects the view that metals are a combination of sulphur and mercury because he knows that the combination of these two substances yields cinnabar, which is not a metal. Jungius also considers whether each compound substance must contain all simple bodies.

A more precise critique of both the Aristotelian (peripatetic) and alchemical (spagyric) formulation of the element was provided by Robert Boyle in 'The Sceptical Chymist' (1661). It seems that Boyle was familiar with Jungius's view. Boyle's work was one of the milestones in the development of chemical ideas. In this treatise, three interlocutors: Carneades—who represents Boyle's views and doubts: Themistius—an adherent of Aristotelian or peripatetic theories; and Philoponus—an advocate of the alchemists or spagyrics (who are also called chemists by Boyle) are trying to convince the fourth interlocutor Eleutrius about the correctness of their views. In his book Boyle concentrates mainly on the problems of the number of the elements, their influence on the properties of bodies, and the mechanism of the effect of fire. We shall now discuss the first two problems, the third will be presented in Chapter 5. As a starting point we shall quote Carneades' introductory argument:

Notwithstanding the subtile reasonings I have met with in the books of the peripatetics, and the pretty experiments that have been shewed me in the

Fig. 8. Robert Boyle, acc. to [8].

laboratories of chymists, I am of so diffident or dull a nature, as to think that if neither of them can bring more cogent arguments to evince the truth of their assertion than are wont to be brought a man may rationally enough retain some doubts concerning the very number of those material ingredients of mixt bodies, which some would have us call elements, and others principles. Indeed when I considered that the tenets concerning the elements are as considerable amongst the doctrines of natural philosophy, as the elements themselves are among the bodies of the universe, I expected to find those opinions solidly established upon which so many others are superstructed. But when I took the pains impartially to examine the bodies themselves that are said to result from the blended elements, and to torture them into a confession of their constituent principles, I was quickly induced to think that the number of the elements has been contended about by philosophers with more earnestness than success. (...) I must desire you then to take notice (...) that my present business doth not oblige me so to declare my own opinion on the subject in question as to assert or deny the truth either of the peripatetic or the chymical doctrine concerning the number of the elements, but only to shew you that neither of these doctrines hath been satisfactorily proved by the arguments commonly alledged on its behalfe. (...) and as have hitherto been wont to be brought either to prove that 'tis the four peripatetic elements, or that 'tis the three chymical principles that all compounded bodies consist of [32, pp. 15–16].

Thus, Boyle asks three questions: whether

(1) substances regarded by peripatetics and spagyrics as simple are really simple;

(2) the statement that each compound substance contains all simple substances is correct;

(3) the number of simple substances is really limited to four, as follows from what the Aristotelians have claimed, or to three, as the alchemists suggest.

Let us see what arguments Boyle uses to give negative answers to all three questions.

The views of the Aristotelians are expressed by Themistius who seemingly upholds them. At the outset Themistius emphasizes that the theory of the four elements is in accordance with the general philosophy of nature, and next he asserts:

(...) I must proceed to tell you, that though the assertors of the four elements value reason so highly, and are furnished with arguments enough drawn from thence, to be satisfied that there must be four elements, though no man have ever yet made any sensible trial to discover their number, yet they are not destitute of experience to satisfie others that are wont to be more swayed by their senses than by their reason. And I shall proceed to consider the testimony of experience, when I shall have first advertised you, that if men were as perfectly rational as 'tis to be wished they were, this sensible way of probation would be as needles as 'tis wont to be imperfect. For it is much more high and philosophical to discover things *a priore* than *a posteriore*. And therefore the peripatetics have not been very solicitous to gather experiments to prove their doctrines, contenting themselves with a few only, to satisfy those that are not capable of a nobler conviction. And indeed they employ experiments rather to illustrate than to demonstrate their doctrines, (...) [32, p. 20].

2. THE ELEMENT

Themistius continues to describe a piece of wood burning in a chimney and carries an analysis of observation analogous to that quoted earlier. Next he proceeds with his argument:

For this (peripatetic) doctrine is very different from the whimseys of chymists and other modern innovators, of whose hypotheses we may observe, (...) that as they are hastily formed, so they are commonly short-lived. For so these, as they are often framed in one week, are perhaps thought fit to be laughed at the next; and being built perchance but upon two or three experiments are destroyed by a third or fourth, whereas the doctrine of the four elements was framed by Aristotle after he had leasurely considered those theories of former philosophers (...). Nor has an hypothesis, so deliberately and maturely established, been called in question till in the last century Paracelsus and some few other sooty empirics, rather than (as they are fain to call themselves) philosophers, having their eyes darkened, and their braines troubled with the smoak of their own furnaces, began to rail at the peripatetic doctrine, which they were too illiterate to understand, and to tell the credulous world that they could see but three ingredients in mixt bodies [32, pp. 21–22].

As a matter of fact Themistius seems to argue that reasoning is more valid than experiment.

As we have seen in Chapter 1, in Boyle's time this statement, derived from the Eleatics, was no longer commonly accepted, and therefore Boyle sounds deliberately ironical here. He also adds his own comment in the form of Carneades' repartee to the ridiculed views:

And first (...) I might here make a great question of the very way of probation which he and others employ, without the least scruple, to evince that the bodies commonly called mixed are made up of earth, air, water, and fire, which they are pleased also to call elements; namely that upon the supposed analysis made by fire, of the former sort of concretes, there are wont to emerge bodies resembling those which they take for the elements. For (...) I might alledge, that by Themistius his experiment it would appear rather than those he calls elements are made of those he calls mixt bodies, than mixt bodies of the elements (...). Nor has any man (...) yet proved that nothing can be obtained from a body by the fire that was not pre-existent in it [32, pp. 23, 24].

Boyle also criticizes the alchemists' views; his alter ego Carneades emphasizes that nobody has ever proved how many bodies the alchemists should really accept as elements. They were quite willing to regard water and earth as elements too. Furthermore, Boyle/Carneades asserts that for the peripatetics only these two bodies are

2.3. CONCEPT OF THE ELEMENT

elements, because fire above air "being generally by judicious men exploded as an imaginary thing; and the air (...) only lodging in pores" (of bodies—R.M.) [32, p. 187].

Let us examine in more detail the critique of the spagyric elements carried out by Boyle/Carneades:

> I told you already that there is a great difference betwixt the being able to make experiments, and the being able to give a philosophical account of them. (...) But that which I would rather have here observed is, that the chymists I am now in debate with have given up the liberty (...) of using names at pleasure, and confined themselves by their descriptions, (...) so that although they might freely have called anything their analysis present them with, either sulphur, or mercury, or gas, or blas, or what they pleased; yet when they have told me that sulphur (for instance) is a primogeneal and simple body, inflammable, odorous, etc. they must give me leave to disbelieve them, if they tell me that a body that is either compounded or uninflammable is such a sulphur; and to think they play with words, when they teach that gold or some other minerals abound with an incombustible sulphur, which is as proper an expression, as a sun-shine night, or fluid ice [32, p. 117].

Boyle does not deny that some bodies can be referred to as salt, mercury, or sulphur, but he doubts whether these substances are simple bodies:

> (...) that though chymists pretend from some (minerals) to draw salt, from others running mercury, and from others a sulphur; yet they have not hitherto taught us by any way in use among them to separate any one principle, whether salt, sulphur, or mercury, from all sorts of minerals without exception. And thence I may be allowed to conclude that there is not any of elements that is an ingredient of all bodies, since there are some of which it is not so [32, p. 196].

> And here I further observe, that I never could see any earth or water, properly so called, separated from either gold or silver (...) and I may conclude, that (...) there is neither earth nor water; I may be allowed to conclude, that neither of those two is an universal ingredient of all those bodies that are counted perfectly mixt [32, p. 197].

Thus, Boyle does not question the results of the experiments made by the alchemists, but he rejects their reasoning. He asserts explicitly:

> (...) chymists have been much more happy in finding experiments than the causes of them; or in assigning the principles by which they may best be explained [32, p. 227].

2. THE ELEMENT

He repeatedly asserts that neither the peripatetic elements nor spagyric elements are simple substances. Thus one could presume, as Eleutrius suggests, that simple substances do not exist at all. Carneades/Boyle's repartee to such a supposition contains a completely new conception of a chemical element. Boyle states that there exist bodies in nature which must be regarded as elements:

Now the considerations that induce men to think, that there are elements, may be conveniently enough referred to two heads. Namely, the one, that it is necessary that nature make use of elements to constitute the bodies that are reputed mixt. And the other, that the resolution of such bodies manifests that nature had compounded them of elementary ones [32, p. 188].

Previously he defines the term element:

I now mean by elements (...) certain primitive and simple, or perfectly unmingled bodies; which not being made of any other bodies, or of one another, are the ingredients of which all those called perfectly mixt bodies are immediately compounded, and into which they are ultimately resolved: now whether there be any one such body to be constantly met with in all, and each, of those are said to be elemented bodies, is the thing I now question [32, p. 187].

Carneades/Boyle strongly emphasizes the following definition:

It may likewise be granted, that those distinct substances, which concretes generally either afford or are made up of, may without very much inconvenience be called the elements or principles of them [32, p. 34].

These two definitions have become perhaps the most revolutionary formulations of the concept of the 'element'. Both the Aristotelian and alchemical elements were, as we have seen, carriers of certain properties. In Boyle, the concept of the element appeared as the end of analysis. This concept of the element functioned in an almost unchanged form for nearly two hundred years. Lavoisier understood the word 'element' likewise. Similarly, in the third edition of his textbook *Początki chemii* (The Principles of Chemistry), Jędrzej Śniadecki wrote:

(...) and the terms of elements or simple bodies (...) will signify rather the final decompositions than the final natural entities [51, Vol. I, p. 3].

Similar formulations can be found in the chemical textbooks of the 1860s.

2.3. CONCEPT OF THE ELEMENT

If simple bodies do exist in nature, a question arises: how many? In this book Boyle considers this problem several times:

(...) it does not appear that three is precisely and universally the number of the distinct substances or elements, wherein to mixt bodies are resoluble by the fire [32, p. 95]. (...) it seems very possible, that to the constitution of one sort of mixt bodies two kinds of elementary ones may suffice, another sort of mixts may be composed of three elements, another of four, another of five, and another perhaps of many more (...) it being very probable that some concretes consist of fewer, some of more elements [32, p. 97]. (...) so much conduceth to the perfection of the universe, that all elemented bodies be compounded of the same number of elements, than it would be for a language, that all its words should consist of the same number of letters [32, p. 185].

Considering the components of plants and animals Boyle repeatedly refers to the works of Paracelsus, according to which

(...) mineral bodies (...) may be resolved into a saline; a sulphureous, and a mercurial part (...) almost all vegetable and animal concretes may (...) be divided into five differing substances: salt, spirit, *oyle*, phlegme and earth [32, p. 228].

This is another confirmation of Carneades' final conclusion

(...) that there is not any certain determinate number of such principles or elements to be met with universally in all mixt bodies [32, p. 226].

Of a particular significance for the development of the concept of the element was Boyle's critical examination of a view which claimed that in composite bodies the elements still manifested their individual properties. The starting point for this criticism was Sennert's statement: *Ubicunque (saies he) pluribus eadem affectiones et qualitates insunt, per commune quoddam principium insint necesse est, sicut omnia sunt gravia propter terram, calida propter ignem* [32, p. 166].

(Wherever the same effects and interactions exist, it is necessary that certain elements have common properties; for example, all the heavy ones are caused by the earth and the hot ones by the fire.)

Carneades/Boyle argues against the above statement:

(...) how can he prove, that the gravity of all bodies proceeds from what they participate of the element of earth? Since we see, that nor only common water, but the more pure distilled rain water is heavy; and quicksilver is much heavier than earth itself; though none of my adversaries has yet proved, that

58   2. THE ELEMENT

it contains any of that element. And I rather make use of this example of quicksilver, because (...) if it will be demanded how it comes to be fluid, they (the peripatetics—R.M.) will answer, that it participates much of the nature of water [32, p. 167].

Next, the author points out that it is a nonsense to claim that mercury consists of earth, water, and air, because they are all lighter than mercury. Similarly weight is not a property given by any of the elements; nor is it analogously liquidity or ability to move or luminosity. Thus Carneades/Boyle concludes:

But I consider further, that chymists are far from being able to explicate by any of the *tria prima* (three spagyric elements—R.M.) those qualities which they pretend to belong primarily unto it, and in mixt bodies to deduce from it. 'Tis true indeed, that such qualities are not explicable by the four (Aristotelian—R.M.) elements [32, p. 175].

Boyle sought for the causes of different properties of compounds, which he called mixed bodies, in their structure and not only in their composition. This problem will be discussed more extensively in Chapter 4.

We can see from the fragments quoted above that Jungius and Boyle understood the concept of the element almost in its modern meaning. Boyle expressed it in a more explicit way. He had strong and justified doubts whether any of the Aristotelian or alchemical concepts satisfies the conditions which an element should satisfy; he questioned the recognized numbers of the elements. Unfortunately his whole treatise 'The Sceptical Chymist' has merely a sceptical/critical character; it lacks explicitly constructive argumentation. Boyle rejects the view that substances held as elements by his contemporary scientists are actually elements; however, he does not mention any substance which should be considered as an element. Carneades/Boyle is not even certain if

(...) gold, which is a body so fixt, and wherein the elementary ingredients (if it have any) are so firmly united to each other, that we finde not in operations wherein gold is exposed to the fire (...) it does discernably so much as lose of its fixedness or weight [32, p. 39].

This lack of constructive argumentation probably delayed general acceptance of Boyle's views about the elements although his experi-

2.3. CONCEPT OF THE ELEMENT

ments and opinions were well known and discussed by the fellow members of the Royal Society.

In 1669, Johann Joachim Becher, who is still in favour of Aristotle's four elements, states that one of them—earth—occurs in three forms which correspond as a matter of fact to the spagyric elements;

(...) I assert that three different earths are contained in metals and rocks; the first, apart from its mixes is found in rocks or in alkaline salt; the second being in saltpetre; and the third in common salt; therefore when mixed without any other ingredient the three earths form a real elemental metal and a rock susceptible to transmutation; of which I conclude that both metals and rocks are also composed of them in a natural way [52, p. 118].

Next Becher discusses the properties of the three earths:

The first principle of metals and rocks, which is glassy or rocky earth (*terra lapida*) and inaccurately called salt (...). Without it no mineral has any value (...). Therefore, we consider this earth to be the primary principle of all metals, minerals, rocks, and crystals [52, pp. 119–121].

The second principle of minerals, which is fatty earth (*terra pinguis*) and inaccurately called sulphur (...). This fatty earth related to another earth, of which we have already spoken, and from which metals receive liquidity and meltability, is certainly much related and similar to the earth of plants (...) which is present during the calcination and leaching of ash [52, p. 132].

In the texts quoted earlier 'fattiness' of sulphur has been mentioned. Jungius attributes ability to combust simply to the 'fattiness' of substances. This term should be understood as 'greasiness'.

The third principle of minerals, which is a liquid earth (*terra fluida*) and inaccurately called mercury (...). What makes metal metallic is derive from this third principle. I have to conclude that this third principle is called mercury due to its mercurial property, i.e. volatility, because when attacked by this earth metal becomes again liquid and volatile [52, pp. 149, 152].

Of course, what Becher has in mind here is the amalgamation of metals by mercury.

In the late 17th century the most popular presentation of chemistry was a systematic discourse entitled *Cours de chymie*, written by Nicolas Lemery, first published in 1675 and reprinted a number of times. According to the foreword to the 1697 edition, its first publication was "a best seller like a work of wit and satire". A systematic review of the properties of mineral, plant, and animal substances was preceded by Lemery's consideration on chemical laws and principles:

The first principle which can be assumed as a constituent of compound bodies is a certain universal spirit which being spread everywhere produces different substances depending on various matrices and pores of the earth within which it is confined. As the chemists who performed an analysis of different compound bodies have found five strong substances and came to a conclusion that there exist five principles of natural substances: water, spirit, oil, salt, and earth. Of these five, three are active: spirit, oil, and salt; and two are passive: water and earth. They are called active because being in a great motion they cause a compound body to act. The remaining ones are called passive because being at rest they serve to retain the vivacity of active substances.

The spirit, called Mercurius, is a subtle, permeable, and light substance which is most effervescent of any other principles. It causes formation of compound bodies in a shorter or longer time depending upon the fact whether it meets a small or large amount, and also due to its strong motion.

Oil, which is called sulphur because of this inflammability, is a mild, subtly greasy substance which escapes from the body following the spirit. It is believed that it contributes to the variability of colours and scents.

Salt is the heaviest of the active principles (...) it gives weight to compound bodies (...). The salt of compound bodies is divided into three kinds: fixed salt, volatile salt, and essential salt.

Water, which is called phlegm, is the first of the passive principles. During distillation it escapes before the spirits, when they are fixed, or after them (...). It serves to diffuse the active principles and to weaken their action.

Earth called a 'dead head' (in Latin *caput mortuum*—this term signifies the deposit which remaines after distillation—R.M.) is the last of the passive principles (...).

The name of principles in chemistry ought not to be treated precisely because substances being thus referred to are principles only from our point of view, (...), because we cannot go further in the division of bodies, but it is well understood that these principles are further divisible into infinitely many parts which deserve even more to be called principles. Therefore in accordance with our weak efforts only isolated and separated substances are assumed in chemistry to be principles. And since chemistry is a science based on demonstrations, it takes for granted only what can be touched and shown [53].

As can be seen, Lemery's views which, many years after the publication of Boyle's 'The Sceptical Chymist', exerted an influence on the reasoning of chemists, were based on the existence of spagyric

2.3. CONCEPT OF THE ELEMENT

elements, which Lemery, like some alchemists, attempted to link with Aristotle's elements.

Thus, the view that earth, water, and air are the elements was still upheld by the 18th century chemists. As late as 1777, Karl Wilhelm Scheele stated that the "majority of chemists seem to be attached to the peripatetic elements". It is greatly to Lavoisier's credit that he ultimately deprived the Aristotelian elements/principles of the rank of elements and carried out a quantitative proof that earth, water, and air are in reality complex bodies. However, before we show this, we must first turn to the experiments and considerations related to the problem of changes in the weight of bodies caused by the action of fire.

As has been mentioned, in 1630 Rey attributed an increase in the weight of tin and lead, observed as a result of roasting, to an increase in the weight of air contained in the pores of metal. On the other hand, Boyle attributes an increase in the weight of metal to the attachment of the particles of fire:

(...) if that be true which was the opinion of Leucippus, Democritus, and other prime atomists of old, and is in our dayes revived by no mean philosophers; namely, that our culinary fire, such as chymists use, consists of swarmes of little bodies swiftly moving, which by their smallness and motion are able to permeate the sollidest and compactest bodies, and even glass itself; if this (I say) be true, (...) it will not be irrational to conjecture, that multitudes of these fiery corpuscles, getting in at the pores of the glass, may associate themselves with the parts of the mixt body whereon they work, and with them constitute new kinds of compound bodies [32, p. 119].

A similar view is presented by Lemery in his textbook when he discusses an increase in the weight of lead while heating:

Lead is a metal saturated with sulphur. During the calcination of lead and many other materials there occurs an effect which is worth consideration, i.e. although under the action of fire the sulphurous or volatile part of lead is dissipated, which should decrease its weight; however, after a long calcination it is found to weigh more and not less.

Some contemporary investigators explained this phenomenon by the attachment of part of the burnt fuel to metal, but the French chemist was of a different opinion:

This effect can be accounted for rather by the fact that the pores of lead are arranged in a way that if the corpuscles of fire penetrate them, they will remain bonded and glued in the pores, attached to the metal and hampered without a way out and thereby they increase weight. I know that I shall be reproached since the corpuscles of fire being very light by nature cannot so much increase the weight of lead. I believe, however, that a very great number of them enters the pores of the metal [53, XI ed., p. 92].

As a matter of fact, the considerations of Rey, Boyle, Boerhaave, Lomonosov, and Lemery related to an increase in the weight of metals under the influence of heating, concern not so much the mechanism of combustion, as it is often interpreted, but rather the question whether fire has some weight. The 17th century scientists, who held fire to be an element, must have considered this problem after accepting Galileo's hypothesis that all the elements have weight, some smaller, others greater. They could have solved this problem only by applying the law of conservation of the mass of the elements, which had not yet been formulated at that time.

Let us note another explanation of an increase in the weight of metals in the process of heating by the attachment of some of its parts, which preceded Lavoisier's explanation of the burning process. It is contained in a textbook of physics published in 1745 and written by Abbé Nollet, a member of the French Academy and the Royal Society. After presenting the opinions quoted above about an increase in the weight of metal in the process of roasting Nollet writes:

I am not convinced whether these opinions are so decisive as one could think, because one may wonder if this increase in weight is caused, strictly speaking, by fire or by other matter which is being attached to a calcinated body; it can be derived from air which touches it, or from the vessels in which they are contained, or from instruments with which we mix them during heating, or eventually from coal which we supply to fire (...). Yet, although the experiment has never proved beyond doubt whether fire has weight, one cannot say that it has proved something contrary. Since the scales did not lose their equilibrium when the same body was weighed cold and then hot, it should be rather presumed that an increase in the weight of heated bodies was not high enough to disturb the instrument and not equal to zero. And as all known matters have weight, it should not be presumed that the matter of fire be an exception from the general law without providing positive and obvious proofs [54].

Nollet's views seem to be in conformity with the theory of combustion which Lavoisier advanced several dozen years later. Similar views are propounded by Voltaire in 1752 and J. Black in 1762. In 1774 G. L. de Buffon thinks that air is attached to metal, and in 1776 Erxleben believes that attachment of carbonic acid (CO_2) causes weight to be increased during heating. The problem of weight increase was finally solved by Lavoisier. He based his theory on a few premises derived from experimental research carried out throughout the 18th century and which was related to the theory of phlogiston and the development of pneumatic chemistry, i.e. the chemistry of gases.

2.4. The theory of Phlogiston

J. J. Becher was a forerunner of the phlogiston theory. In the preceding section we have cited his views concerning the existence of chemical principles/elements. In a work published in 1667 he deals, amongst other things, with the process of combustion; however, he does not say explicitly which substance is burned and which substance sustains combusion. For example, he holds that a gas that is given off during the calcination of saltpetre, oxygen, as we know nowadays—also contains a "fatty earth" [55, p. 271]. He asserts that only highly rarefied bodies can be burnt because "in the process of combustion they are dissipated into atoms" [55, p. 219], and contain salty parts, composed in a great measure of water "changing to air during combustion" [55, p. 278]. He also asserts that "combustion, i.e. calcination by a strong fire, acidifies bodies because combustion is analogous to calcination, although the latter lasts longer" [55, p. 135]. Incineration, fermentation and calcination are analogous processes, but the substances of the 'kingdom of minerals" (metals) are calcinated, and the substances of the kingdom of plants are fermented prior to burning.

The process of fermentation and combustion was analysed more precisely by an adherent of Becher's views, Georg Ernest Stahl in his four works published in the years 1697, 1716, 1738, and 1747. Stahl attributes combustibility to the presence of the peculiar matter of fire in bodies, the Becherian principle—sulphur. In his work of 1716 he writes: "Toward fire, this sulphur principle behaves in such a manner that it is not only suitable for the movement of fire but is also one and the same being, yes, even created and designed for it. But also, according to a reasonable manner of speaking, it is the corporeal fire, the essential fire material, the true basis of fire movement in all inflammable compounds. However, except in compounds, no fire at all occurs, but it dissipates and volatilizes in invisible particles, or at least, develops and forms a finely divided and invisible fire, namely, heat."

Next Stahl emphasizes that this principle, i.e. the 'matter of fire' occurs only in a combined form with other substances, and never in the free state. He writes:

From all these various conditions, therefore, I have believed that it should be given a name, as the first, unique, basic, inflammable principle. But since it cannot, until this hour, be found by itself (...), outside of all compounds and unions with other materials, and so there are no grounds or basis for giving a descriptive name based on properties, I have felt that it is most fitting to name it from its general action, which it customarily shows in all its compounds. And therefore I have chosen the Greek name *phlogiston*, in German *brennlich* (inflammable) ... [56].

Stahl was not the first to use this Greek word. Aristotle used it in the sense of 'heat'; and at the turn of the 16th century, Nicolaus Niger Hapelius employed this term to denote combustibility, while Sennert—according to Boyle—held combustibility–phlogiston—to be one of the properties of substances. Stahl was probably familiar with the terms used by Hapelius and Sennert, which may have induced him to use this verb. However, it was only Stahl's 'phlogiston' which proved to be important in the evolution of chemistry.

Stahl held that phlogiston was contained above all in substances belonging to the kingdom of plants and animals, mostly in fats. On the other hand:

In the mineral kingdom there is nothing but water, common salt, pure vitriolic salts, and stones in which the substance (i.e. phlogiston—R.M.) is little or not all found. On the other hand, coal and bitumen are full of it; sulphur, not indeed in weight, but in the number of its finest particles, is completely possessed with it. Not less is it found in all inflammable, incomplete, and so-called 'unripe' metals [57].

During a chemical transformation phlogiston, he assumed, passes

2.4. THE THEORY OF PHLOGISTON

from one substance to the other. In the first of the cited works Stahl writes explicitly:

Of course I can indicate many experiments to show how phlogiston passes directly from fats or carbon to the metals themselves and regenerates them from burnt lime (ores—R.M.) to a malleable and amalgamable form [47].

And next: "The instability of metals consists in the fact that their essential fire material is easily taken up by air" [57, p. 391], but "a burnt metal can regain its metallic form when given such a matter of fire" [58].

A description of his experiment with sulphates is also frequently quoted: "...when calcinated with carbon in air these clusters yield sulphate liver (a mixture of sulphides and polysulphides—R.M.), and water, whereas this mixture, after more calcination with carbon gives off sulphur and solid alkaline salt" (i.e. sodium hydroxide—R.M.).

When Stahl calcinated pure sulphur with solid alkaline salt, he obtained sulphate liver which, when heated in an open crucible (i.e. with access of air) was transformed into sulphates. Stahl asserts that "this cycle can be repeated a hundred times" [47, p. 182].

He accounts for the appearance of sulphur in the above cycle by the fact that vitriolic acid (i.e. SO_2 given off from the sulphates) "takes the fire material from carbon and then forms a real sulphur" [47, p. 181].

The above example of a transformation cycle: sulphur–sulphides–sulphates shows that the phlogiston theory qualitatively arranged the successive stages of calcination of a given substance; today we would refer to them as the oxidation states of a given element.

Dissolution of metals in nitric acid is accounted for by Stahl in the following way: in the presence of metals saltpetre becomes susceptible to the reception of the fire material, i.e. phlogiston which is contained in metals, while humidity which is present in the fire material dissolves metal. Lead is dissolved less readily than iron because it contains less phlogiston [47, p. 254].

Stahl does not discuss changes in the weight of bodies after combustion. In the year 1723, using the term *Brennliches Grundwesen* (the matter of fire) he writes:

(...) however this essential being is not the lightest of beings treated as bodies, namely it is much contained in a mass (i.e. volume—R.M.) but little in weight. Therefore, when accessed by bodies, (...) they become lighter and more loose (i.e. have more voids—R.M.) [59, p. 70].

In another work he states that bodies will become either lighter or heavier under a strong effect of fire [58, p. 307].

Thus, the phlogiston content in a substance predetermines, in Stahl's opinion, the possibility of its reaction with another substance, particularly with one that lacks the fire material. A substance that was believed to contain little phlogiston was nitric acid. This reasoning provided a basis for developing the chamber process of obtaining sulphuric acid whereby sulphur dioxide oxidizes to sulphur trioxide by reaction with nitric oxides, and next dissolves in water producing sulphuric acid. Also, the Leblanc method of obtaining soda by melting sodium sulphate, obtained in consequence of subjecting table salt to the action of sulphuric acid, with carbon and calcium carbonate was developed on the basis of the concept of phlogiston.

As we can see, Stahl took no account of the argumentation which Boyle had introduced a few dozen years earlier. Although Stahl did not take into consideration the criticism of the peripatetic and spagyric elements previously made by Boyle, his theory (i.e. Stahl's) is closely based, in contrast to the earlier theories, on the observed facts, and does not contain any mystical elements. The element phlogiston, identified with the spagyric sulphur and the Aristotelian fire, was believed to be—just like other spagyric elements—the carrier of one property, in this case combustibility.

We have already noted that the phlogiston theory introduced some arrangement to reactions which are today called redox reactions. In 1766, Cavendish identified phlogiston with the inflammable air, i.e. hydrogen, which he had just discovered. In 1819 T. C. Grothus again identified phlogiston with negative electrical charges and the modern historians of science, J. Bernal [2] and F. Szabadvary find a certain analogy between phlogiston and electrons. Of course, it would be inadmissible to claim that the phlogisticians had anticipated the existence of electrons, but this comparison shows that an explanation of the mechanism of combustion by the transmission of some property

2.4. THE THEORY OF PHLOGISTON

is not absurd. The introduction of some order to the views on the reactions of oxidation/reduction enabled the adherents of the concept of phlogiston to predict the course of unknown reactions of the same kind, e.g. that of nitric acid with sulphur dioxide. It should be noted that, thanks to the phlogiston theory, it was possible for the first time to plan and devise new technological processes, some of which were of essential significance for the development of the chemical industry, such as the methods of obtaining sulphuric acid and soda.

The eminent British historian of chemistry James Partington [16] emphasizes that at the end of the 17th century, chemists were no longer satisfied with a mechanistic approach to chemical processes, which resulted in a rebirth of some alchemical conceptions. However, on the other hand, we have noted that the study of the processes involved in the calcination of ores was imposed by the needs of metallurgy. Although Stahl's theory was based on fallacious premises, it enabled chemists to give metallurgists some effective advice. Therefore, the phlogiston theory was the first theory which exerted a significant influence on the development of technology.

This theory, as we have said, combined the phenomenon of heat with the phenomenon of combustibility. It was the first concept which treated uniformly the result of calcination (roasting) and combustion in a flame. In both processes the substance was believed to lose the property of combustibility and thus to liberate phlogiston. The view that a composite body is decomposed in fire into its constituents agreed with the contemporary opinion, which was derived from Aristotle, that fire purifies bodies. This notion was so deeply rooted among contemporary scientists that Boyle's arguments passed unnoticed. (We shall return to this problem in Chapter 5.) It was commonly held that phlogiston is liberated from a burning body which in the process of calcination loses its combustibility and ceases to change. A noninflammable body which does not change under the influence of heat cannot contain phlogiston.

The above argumentation also referred to metals. This is not surprising; after all Stahl identified phlogiston with philosophical sulphur, and—despite Boyle's arguments—it was still commonly believed that philosophical sulphur, i.e. phlogiston according to the new ideas, was a constituent of metals, which were not yet regarded as elements. Besides sulphur–phlogiston, metals were believed to include earth. In the 18th century, it was already accepted that there are several kinds of earth. Thus metal was held to consist of earth and phlogiston. In the process of calcination, phlogiston escapes from metal, leaving earth behind.

However, it was already known from the quantitative investigations that we have discussed earlier that metal becomes heavier in consequence of calcination. We have said that Boyle and Lemery accounted for this increase in weight by the attachment of the weighty particles of fire. Nonetheless, phlogiston was given off, and it was not attached to substances; in order to resolve this dilemma some scientists in the 18th century, e.g. Joseph Black, stated that phlogiston does not gravitate to the Earth's centre but tends to rise into the air. In the years 1768–1769, this view was stated more precisely by J. G. Wallerius and J. P. Charderon. The latter writes:

Weight is an essential property of matter but lightness is not (...) Particles in a compound less dense than air will tend to rise, and if they cannot leave the compound they will destroy a quantity of the gravitation of the others, proportional to the excess of their lightness to air [60].

Finally, in 1772 Guyton de Morveau was quite explicit about this:

Phlogiston has not an absolute negative weight but is lighter than air; so that a body containing it is buoyed up in the same way as a piece of lead in water weighs less when a cork is attached to it [61].

Let us note, however, that—based on what we know from the available fragments of the original works—none of the scientists stated explicitly that phlogiston has a negative weight, although such a supposition can be found in many later publications on the subject. It was maintained only that phlogiston rises into the air, hence came the conclusion that it must be lighter than air. It is scarcely probable that in the mid-18th century, marked by Newton's authority, a serious scientist would have admitted the feasibility of a matter having negative weight.

As late as 1789, S. F. Hermbstädt [62] attempts to include in the reaction balance not only substances but also the matter of heat, i.e. caloric, and phlogiston. He states that sulphur consists of its principle and phlogiston; and air must also contain its principle and caloric because he treats gases as solutions of solid bodies in caloric. In the process of sulphur combustion, phlogiston, being released from it, combines with air, and the caloric of air, combining with sulphur, causes a gaseous sulphuric acid (SO_2) to be formed, and in part it gives light and heat.

The true interpretation of the phenomena of combustion, which rejected the existence of phlogiston, was provided by Lavoisier when more data were obtained from the investigation of gases—in the years 1756–1774.

2.5. The Discovery of Gases

Greek science distinguished only two kinds of gases: 'air' regarded as an element, and 'spirits' liberated in the process of torrifying of substances. In the early Middle Ages it was found that the vapours, after being dissolved in water, have the properties of acids. Thus these vapours had been called acids for many centuries. Asphyxiating air was known to exist in a grotto near Naples. From ancient times a reducing or oxidizing atmosphere was created in the process of the manufacture of stained glass or decorative ceramics. It was therefore known that air does not always have the same properties. Moreover, as Bugaj [63] has pointed out, as early as the early 17th century. Sendigovius realized that thanks to one of its properties air is indispensable for life. He also knew that this property is increased when saltpetre is roasted in air; therefore the 'spirit of saltpetre' had the property of sustaining life. However, Sendigovius did not write explicitly that he managed to obtain the 'spirits' in a pure state. It was not yet possible in the 17th century to collect substances which could be liberated in a gaseous state.

Van Helmont was the first to realize the existence of different kinds of 'air'. In 1648, he stated that the 'spirit' obtained from the combustion of wood suppresses flames, in contradistinction to ordinary air. Here is what van Helmont wrote about that spirit:

2. THE ELEMENT

I call this spirit, unknown hitherto, by the new name of gas, which can neither be contained by vessels nor reduced into a visible body [49, p. 106].

Van Helmont introduced the word 'gas' as the Dutch pronunciation of this word was 'chaos'; the scientist wanted to emphasize that a gaseous state is the chaotic state of matter. Van Helmont called the gas which he obtained by combustion, *gas silvestris* (carbon dioxide).

The application of a pneumatic tank, in 1724, enabled pure gases to be collected and stored for quantitative experiments. In 1756, i.e. a hundred years after gas silvestris was discovered, Joseph Black obtained the same gas by heating limestone. Van Helmont's discovery had already been forgotten, and because Black obtained this kind of 'air' from a solid body, he called it 'fixed air' (Latin *aer fixus*).

Another gas which was examined in the second half of the 18th century was liberated during the reaction of acids with some metals. It had been known previously because effervescence, accompanying a gas given off by a substance poured on metals and some salts, was the principal technique for identifying acids until the time of Boyle. However, it was not until the year 1766 that Sir Henry Cavendish, proved that it was a combustible gas, much lighter than air; and on this basis he identified that 'inflammable air' with phlogiston.

The years that followed brought the discovery of two more kinds of 'air'. In 1772, a pupil of Blacks', Daniel Rutherford examined the air in which a mouse was kept until suffocation, following which, after the extinction of a candle flame and burning phosphorus, he removed 'fixed air' by calcium oxide, and found that nothing could burn in such air. Since, according to the phlogiston theory, phlogiston is given off in the process of burning, he came to a conclusion that the air which remained in the vessel after burning had been phlogisticated to such a degree that it could not absorb any more. Therefore, he called it 'phlogisticated air'. Rutherford's experiments provided a qualitative proof that air is not a simple substance but constitutes a mixture of at least two gases.

In 1775, Joseph Priestley announced the result of his experiments with air extracted from *mercurius calcinatus* (mercuric oxide):

2.5. THE DISCOVERY OF GASES

Having got about three or four times as much as the bulk of my materials, I admitted water to it, and found that it was not imbibed by it. But what surprised me more than I can well express, was, that a candle burned in this air with a remarkably vigorous flame [64].

Priestley also obtained air of similar properties as a result of heating minium (lead oxide) and discovered that a mouse could live in it. "By this I was confirmed in my conclusion, that the air extracted from *mercurius calcinatus* was at least as good as common air". A further investigation showed that it was even better than common air:

Being now fully satisfied with respect to the nature of this new species of air, viz. that being capable of taking more phlogiston from nitrous air, it therefore originally contains less of this principle; my next inquiry was, by what means it comes to be so pure, or philosophically speaking, to be so much dephlogisticated.

The above considerations prompted Priestley to write the following commentary:

There are, I believe, very few maxims in philosophy that have laid firmer hold upon the mind, than that air, meaning atmospherical air (...) is a simple elementary substance, indestructible, and unalterable, at least as much so as water is supposed to be. In the course of my inquiries, I was, however, soon satisfied that atmospherical air is not an unalterable thing; for that the phlogiston with which it becomes loaded from bodies burning in it, and animals breathing it, and various other chemical processes, so far alters and depraves it, as to render it altogether unfit for inflammation, and respiration (...). (...) from the experiments which I made (...) I was led to conclude, that common air consisted of some acid (...) and phlogiston (...) and therefore that it is not an elementary substance, but a composition.

A few investigators before Priestley had already realized that under certain conditions air could be 'better', i.e. one could breathe in it more easily and bodies could burn more readily. We have already written about the experiments of the Polish alchemist Sendigovius carried out about the year 1600. In 1665, Robert Hooke wrote in his work *Micrographia* that air contains a substance similar to that found in saltpetre. Although Hooke did not know the nature of this substance, he realized that it played a significant role in the process of burning. Similarly, four years later John Mayow mentioned a combination of the 'saltpetre ingredients of air' with metals. He also put a burning candle or a live mouse under a glass bowl immersed in a vessel with water and found that after the loss of about one quarter of the air volume the candle was extinguished and the mouse died. None of these investigators, however, was able to separate from air this constituent which gives life and sustains combustion.

It was Priestley who for the first time isolated one of the constituents of air in pure state and realized that air is a compound body. A similar conclusion was reached two years later by Karl Wilhelm Scheele on the basis of the results of the combustion of different bodies in air:

I conclude from the presented experiments that air consists of two different fluids, one of which does not exhibit in the least degree the property of attracting phlogiston, whereas the other, which constitutes about one third to one fourth of the whole volume of air, is actually susceptible to such attraction [65].

2.6. Overthrow of the Concept of Phlogiston and of the Aristotelian Elements

Almost at the same time, in the 1770s, Lavoisier also came to a conclusion that there must be two different constituents of atmospheric air because:

(...) the whole of atmospheric air is not in a state fit for breathing; this part, which is healthy, combines with metals during calcination, and this part, which remains after calcination, is a kind of choke-damp unfit for maintaining the breathing of animals and burning of bodies; the principle which unites with metals during calcination which increases their weight, and which is a constituent of the calx is nothing else than the healthiest, and purest, part of air [66].

Finally when discussing the process of combustion, Lavoisier distinguishes four characteristic phenomena:

First Phenomenon. In all combustions the matter of fire or light is evolved.
  Second Phenomenon. Materials may not burn except in a very few kinds of air, or rather, combustion may take place in only a single variety of air; that which Mr. Priestley has named dephlogisticated air and which I name here pure air. (...)
  Third Phenomenon. In all combustion, pure air in which the combustion takes place is destroyed or decomposed and the burning body increases in weight exactly in proportion to the quantity of air destroyed or decomposed.

2.6. OVERTHROW OF THE PHLOGISTON THEORY

Fourth Phenomenon. In all combustion the body which is burned changes into an acid by the addition of the substance which increases its weight. Thus, for example, if sulphur is burned under a bell, the product of the combustion is vitriolic acid; (more precisely: sulphur dioxide called acid by Lavoisier—R.M.).

The calcination of metals follows precisely the same laws, and it is with very good reason that Mr. Macquer considers the process as a slow combustion (...). These different phenomena of the calcination of metals and of combustion are explained in a very nice manner by the hypothesis of Stahl, but it is necessary to suppose with Stahl that the material of fire, phlogiston, is fixed in metals, in sulphur, and in all bodies which are regarded as combustible. Now if we demand of the partisans of the doctrine of Stahl that they prove the existence of the matter of fire in combustible bodies, they necessarily fall into a vicious circle and are obliged to reply that combustible bodies contain the matter of fire because they burn and that they burn because they contain the matter of fire [67].

At first, Lavoisier used the terminology of the phlogiston theory, because no other terminology was then known in chemistry. However, he was critical about this theory; and he treated the particular kinds of air as principles/elements. He believed that atmospheric air consisted of a mixture of two elastic fluids, which he proved not only qualitatively but also quantitatively, which is more important. At the same time he proved quantitatively, with the aid of the balance, that metallic lime, i.e. the principle earth, and in reality ore, consists of metal and pure air. On the basis of quantitative investigations, Lavoisier overthrew conclusively the view that air and earth are simple bodies, i.e. the elements; he did so by a consistent use of the law of conservation of mass as his premise, although he did not prove that law, as we have stressed in Chapter 1.

Investigating the properties of the products of combustion of different compounds, Lavoisier became convinced that many of them exhibit acidic properties after being dissolved in water. He wrote:

I have shown that purest air, which Mr. Priestley called dephlogisticated air, is a constituent of many acids, namely phosphoric acid, sulphuric acid, saltpetric acid.

Hence Lavoisier concludes:

(...) purest air (...) is the constituent principle of acidity; this principle being common to all acids and included in each of them. According to these laws

I should call dephlogisticated air—the acid-forming principle or if denoted by the Greek words—the oxygenic principle [68].

Thus, in 1777 Lavoisier recognized the oxygen principle as an element which can be a constituent of different substances. It is worth noticing that this very element seemed to be an alchemical progeny—it was assumed to be a carrier of one property—here acidity.

The Latin name *oxygenium* proposed by Lavoisier for dephlogisticated air soon became widespread and was accepted by all scientists, and ever since has been either accepted literally or adopted in the form of literal translation by almost all languages of the world, e.g. German: *Sauerstoff*, Dutch: *zuurstof*, Russian: *kislorod*. But, for example, in the Polish and Danish languages were introduced in c. 1850 quite original native names derived from the root *tl—tlen* in Polish (Polish *tlen* derived from *tlić* = 'to smoulder') and *ilt* in Danish.

In the course of the investigations into respiration discussed earlier in this book Lavoisier gave a name to that part of air which remained after removing oxygen:

The chemical properties of the noxious portion of atmospheric air being hitherto but little known, we have been satisfied to derive the name of its base from its known quality of killing such animals as are forced to breathe it, giving it the name of azote, from the Greek private article *a* and *zoe* (vita), hence the name of the noxious part of atmospheric air is azote gas [36, Vol. I, p. 67].

The term 'azoth' had already existed in chemistry—it was used by the alchemists to denote Mercury.

Employing consistently as a premise the law of the conservation of mass and the gravimetric method, in 1781 Lavoisier quantitatively resolved water into the oxygenic principle and aqueous principle called in Latin *hydrogenium*. Cavendish, who had discovered this element earlier on the basis of qualitative analysis, in 1784 carried out a quantitative synthesis of water from oxygen and hydrogen.

In his textbook *Traité élémentaire de chimie*, published in 1789, Lavoisier provides quantitative results concerning the decomposition of water which is passed over incandescent coal and incandescent iron. The numerical data given by Lavoisier throw light on the accuracy of his measurements. He employed one grain as a unit of weight—

2.6. OVERTHROW OF THE PHLOGISTON THEORY

which equalled about 0.05 gramme. By passing 85.7 grains (about 4.14 grammes) of water over coal he obtained 100 grains of carbonic acid (i.e. carbon dioxide) and 13.7 grains of hydrogen as a result of the reaction of this amount of water with 28.0 grains of coal. Next,

Fig. 9. Antoine Lavoisier with his wife painted by Louis David, acc. to [8].

having decomposed 100 grains of water with the help of iron, he obtained 15 grains of hydrogen. Water thus contained 85 grains of oxygen [36, Vol. I, p. 91]. Today, knowing the molar masses and atomic weights of compounds and elements, we are able to calculate that in the first experiment he must have used 27.56 grains of carbon for 85.7 grains of water; thus the value given by Lavoisier was correct. And from that amount he must have obtained 103.8 grains of carbon dioxide, which is also in agreement with the amount given in his textbook. However, the amount of hydrogen obtained was much overstated, because Lavoisier should have obtained 9.5 grains, and not 13.7 grains of that gas. On the other hand, in the second experiment 100 grains may have contained 11.11 grains of hydrogen and 88.89 grains of oxygen. According to Lavoisier the quantitative ratio of oxygen to hydrogen, obtained by passing water over incandescent coal was 5.26; and by passing it over incandescent iron was 5.66, Lavoisier gave the amounts of the substances with an accuracy up to 1 per cent. The difference in the results of the analysis of water was 8 per ent. However, Lavoisier did not analyse that difference. Today, it is difficult to say whether he realized the fact that he should have obtained the same weight ratio of the constituents of water—after all the law of constant ratios had not been finally formulated by then—or he may have recognized the results obtained as being in conformity.

So, having quantitatively resolved earth, water, and air into oxygen, nitrogen, hydrogen, and metals, Lavoisier proved conclusively that out of the four Aristotelian elements three of them are in reality mixed bodies. In this way Lavoisier completed the discussion on the peripatetic elements initiated a hundred years earlier by Boyle. The English scientist submitted those elements to a critical examination. Lavoisier, on the other hand, named the substances which ought to be regarded as elementary bodies.

The first obvious outcome of that approach was the publication of his *Reflexion sur phlogiston* in 1783 [69]. The aim of that work was:

to show (...) that all phenomena of combustion and calcination may be explained in a far simpler and easier manner without phlogiston than with it (...) I have reduced all my explanations to a simple principle that pure air, life air, consists of a particular principle which (...) I have called acid-

2.6. OVERTHROW OF THE PHLOGISTON THEORY

forming (...). If we accept this principle, the major difficulties in chemistry will disappear, and all phenomena may be explained with an elementary simplicity. If all this in chemistry is explained satisfactorily without the help of phlogiston, it is then infinitely probable that the latter principle does not exist, it is an imaginary being without foundations; and eventually it is consonant with the principles of sound logic so as not to multiply beings unnecessarily.

Instead of speaking of adding air which had more or less phlogiston, Lavoisier suggested that the processes in question should be treated as the outcome of increasing or diminishing the amount of oxygen. The new conceptions of the mechanism of combustion spread through the scientific world with difficulty. Their author realized that fact:

I do not expect my ideas to be accepted immediately; (...) time will thus confirm or overthrow the views which I have presented.

Priestley, who was the first to separate oxygen, would never admit that it was an element; and Henry Cavendish, who was the first to examine the properties of hydrogen, a year after the publication of Lavoisier's treatise against phlogiston, gave the following description of the results of the investigation into the synthesis of water from oxygen and hydrogen:

(...) we must allow that dephlogisticated air is in reality nothing but dephlogisticated water (...); and that inflammable air is either pure phlogiston, as Dr. Priestley and Mr. Kirwan suppose, or else water united to phlogiston.

Next, Cavendish admits:

There is the utmost reason to think that dephlogisticated and phlogisticated air, as MM. Lavoisier and Scheele suppose, are quite distinct substances, and not differing only in their degree of phlogistication; and that common air is a mixture of the two (...) most other phenomena of nature, seem explicable as well, or nearly as well, upon this (i.e. Lavoisier's hypothesis—R.M.) as upon the commonly believed principle of phlogiston [70].

Cavendish therefore explains the observed phenomena simultaneously with the help of both methods. In his work quoted earlier Cavendish describes how all the hydrogen and one fifth of ordinary air disappeared and mist appeared instead as a result of passing an electric spark through a spherical vessel with a certain amount of hydrogen and ordinary air. He concludes that:

And by this experiment it appears that this dew is plain water, and consequently that almost all the inflammable air and about one-fifth of the common air are turned into pure water.

In the late 18th century hardly anybody made use of the phlogiston theory.

2.7. The Foundations of Chemistry according to Lavoisier

The final overthrow of the Aristotelian elements and rejection of the phlogiston hypothesis made it necessary to formulate the foundations of chemistry anew. This was done by Lavoisier in his textbook *Traité élémentaire de chimie*, published in 1789, which was a synthesis of his research and views. In the preface to that treatise, he defines accurately what he considers to be an element:

It is very remarkable that, notwithstanding the number of philosophical chemists who have supported the doctrine of the four elements, there is not one who has not been led by the evidence of facts to admit a greater number of elements into their theory. (...) I shall therefore only add upon this subject that if by the term elements we mean to express those simple and indivisible molecules of which matter is composed, it is extremely probable we know nothing at all about them; but, if we apply the term elements, or principles of bodies, to express our idea of the last point which analysis is capable of reaching, we must admit, as elements, all the substances into which we are capable, by any means, to reduce bodies by decomposition. Not that we are entitled to affirm that these substances we consider as simple may not be compounded of two, or even a greater number of principles; but, since these principles cannot be separated, or rather since we have not hitherto discovered the means of separating them, they act with regard to us as simple substances and we ought never to suppose them compounded until experiment and observation has proved them to be so [36, XVII].

For Lavoisier, just as for Boyle, the element was the ultimate end of analysis, the former being more explicit about it than any other investigator. By making a consistent use of the law of conservation of mass and quantitative methods, he managed to point out which substances must be recognized as less complex than others. A part will always weight less than the whole; following this reasoning,

2.7. THE FOUNDATIONS OF CHEMISTRY

Lavoisier devised a list of simple substances. On the basis of their chemical and also their physical properties, he divided the elements into four groups (Table 2) [36, V. I, p. 192].

Table 2. Simple substances according to Lavoisier

light	sulphur	antimony	mercury	lime
caloric	phosphorus	arsenic	molybdenum	magnesia
oxygen	carbon	bismuth	nickel	barite
nitrogen	muriatic radical	cobalt	platinum	argilla
hydrogen	fluoric radical	copper	silver	silica
	boron radical	tin	lead	
		gold	tungsten	
		iron	zinc	
		manganese		

An analysis of this list enables us to draw a few interesting conclusions. The most numerous group of the 'elements'—represented in the third column—contains all the then known metals, substances whose oxides have alkaline properties. It should be emphasized that Lavoisier was the first chemist to recognize metals as simple substances. i.e. elements. And since simple substances have no constituents in common with other substances, they cannot be changed into other simple substances using the methods known in the 18th century (and even in the 19th century). It was therefore not until the time of Lavoisier that all attempts at transforming base metals into gold by means of then available methods came to be regarded as a sheer nonsense. For this reason, one should not be surprised that scientists who lived earlier, even ones as outstanding as Boyle or Newton, 'practised alchemy'.

The second group includes substances whose oxides exhibit acidic properties, Lavoisier distinguished three substances which he called 'radicals'. However, he failed to isolate these radicals in the pure state; he knew only hydrochloric, 'fluoric' (hydrofluoric), and boric acids. Nonetheless, he was so deeply convinced that all acids must contain oxygen that he recognized those three acids as oxygen combinations of the hitherto unseparated 'radicals'. However, it was not consistent with his definition of the element given in the introduction to the *Traité*.

The fourth group contains substances which were not yet decomposed at the time of Lavoisier, and which are, as we know today, oxides and hydroxides of some then unknown elements. Dissolved in water, they did not exhibit acidic or basis properties; therefore they could not have been included in the second or third groups.

Simple substances, included in the three groups mentioned above, except for liquid mercury, occur in the solid state. Besides, there was a group of gaseous substances which Lavoisier, also for the first time in the history of chemistry, recognized as elements. Thus, hydrogen, nitrogen, and oxygen were included in the first group of elements. Furthermore, Lavoisier included light and caloric in this group as elements, because it was believed that light and heat were given off just like gases during chemical reactions.

Before we proceed to analyse the premises and consequences of treating light and heat as elements, let us consider Lavoisier's view concerning the different degrees of oxidation; this problem was quite successfully solved by the phlogiston theory. Lavoisier replaced phlogiston by oxygen (*oxygenium*) which be regarded as a carrier of acidity. In accordance with this convention, the stronger the acidic properties in a compound the more acid-forming oxygen it should contain. Lavoisier realized that oxides of non-metals exhibit acidic properties only after they are dissolved in water; therefore, in those solutions any compound of a non-metallic element should be more oxidized than an undissolved compound. Besides, it was already known that sulphur gives two acids, one being weaker and the other stronger. On this basis Lavoisier concluded that the elements which are found in acids must occur in a free state and in four different oxidation states. Lavoisier was not bothered by the fact that he did not know the whole gamut of acids of one element of different oxygen content. He was convinced that his conception was correct, and moreover he was sure that some day these compounds would be discovered (just like the radicals that were unknown to him). He listed them in a special table in his textbook where he arranged the elements and their successive oxidation states which he called *degreés d'oxygenation* [36].

Thus nitrogen in its first oxidation state yielded nitrous oxide, nitrous acid in the second, nitric acid in the third and, in the fourth oxidation state, oxidized nitric acid. Similarly, sulphur yielded: sulphur oxide, sulphurous acid, sulphuric acid, and oxidized sulphuric acid, respectively. 'Oxidized' acids were believed to occur in water solutions, although nobody was able to separate them from these solutions. Lavoisier showed much more imagination in respect of a non-existing muriatic radical. In the first oxidation number this radical was believed to form muriatous oxide, and in the second muriatous acid (both non-existent), in the third muriatic acid, and, in the fourth, 'oxidized muriatic acid' which was simply an aqueous solution of chlorine—a gas separated by Scheele in 1774 which was not regarded as an element by Lavoisier.

Also, some metals could, according to Lavoisier, occur in the four oxidation states. Thus arsenic was believed to yield, respectively: grey arsenic oxide, white arsenic oxide, arsenic acid, and oxidized arsenic acid; while iron yielded: black iron oxide, yellow and red iron oxide, and in the third oxidation state—iron acid.

2.8. The Imponderable Elements

In order to understand why Lavoisier, who rejected all the Aristotelian and alchemical elements, recognized heat and light as simple bodies, we must analyse the views on the nature of heat which prevailed among physicists throughout the 18th century. Together with the spread of the corpuscular theory of matter, heat phenomena began to be interpreted as motion of the particles of matter. The Dutch chemist and physician Herman Boerhaave regarded heat as a manifestation of particle 'trepidation'. His consideration exerted a decisive influence on the views of the Russian scientist Mikhail Lomonosov, who explained heat as the result of a rotary motion of particles, and also on the views of Count Rumford (Benjamin Thompson), who was a superintendent of the Bavarian army from the year 1784. He noticed then that a considerable amount of heat is given off in the process of drilling gun barrels. He thus assumed that the motion of a boring tool (friction) causes the motion of particles. He performed measurements of the heat capacity of a piece of metal and of the shavings removed by the boring tool. In both cases the values of heat capacity were the same, although heat was given off during drilling. Rumford concluded that if heat were a substance, the heat content of the shavings, which lost part of their heat through volatilization, should be smaller. He wrote:

And, in reasoning on this subject, we must not forget to consider that most remarkable circumstance, that the source of the Heat generated by friction, in these experiments, appeared evidently to be inexhaustible.

It is hardly necessary to add, that anything which any isolated body, or system of bodies, can continue to furnish without limitation cannot be a material substance; and it appears to me to be extremely difficult, if not quite impossible, to form any distinct idea of anything capable of being excited and communicated in the manner the Heat was excited and communicated in these experiments, except it be MOTION.

I am very far from pretending to know how, or by what means or mechanical contrivance, that particular kind of motion in bodies which has been supposed to constitute Heat is excited, continued and propagated; ... [71].

Other scientists, including Abbé Nollet and Joseph Black, regarded heat as a special kind of substance. The concept of heat as a substance was genetically linked with the peripatetic fire and spagyric sulphur, as will be shown in Chapter 5. As we have seen the substance of heat was sometimes identified with phlogiston. There were many arguments in favour of either hypothesis. The concept of heat as the motion of particles did not provide any grounds in the 18th century for quantitative measurements of heat phenomena; therefore, this kind of research could then be carried out only with the help of the substantial theory of heat. For this reason in the 18th and the first half of the 19th century this theory was more widespread than the kinetic theory of heat. As a matter of fact J. Herapath writes in 1821: "this, what I call heat, arises from the internal movement of atoms or particles and is proportional to their momentum" [72], but John Herschel, in his classic work 'A Preliminary Discourse on the Study of Natural Philosophy' published in 1830, does not mention the kinetic

2.8. THE IMPONDERABLE ELEMENTS

theory of heat. It is no wonder that Lavoisier, who made use of almost entirely quantitative measurements, relied on the substantial theory of heat, although he treated the matter of fire, i.e. caloric, in a special way. The starting point of his considerations was the existence of different states of aggregation:

> One can say about all natural bodies: they are solid, liquid or in an elastic and aeriform state, depending on the relation between the attractive force of their molecules and the repulsive force of heat, or—which means the same—depending on the degree of heat to which they are exposed.

The idea that heat prevents particles from integrating into one chunk had been expressed a few years earlier by Thorbjön Bergman. As we shall see in Chapter 3, this idea played a significant role in Dalton's considerations. We can learn from the following argument how Lavoisier conceived of the effect of heat:

> It is difficult to conceive of these phenomena without assuming that they are an outcome of the effect of a real and material substance, a very delicate liquid, which penetrates between the molecules of the whole bodies, and which expands them; supposing even that the existence of this liquid is a hypothesis, we shall see that it explains in a satisfactory manner the phenomena of nature. This substance, whatever it might be, is the cause of heat, or in other words, the impression referred to as heat is an outcome of aggregation of this substance. Thus it cannot be called heat in a precise language because the same name cannot be referred to both.

Therefore, together with many other physicists Lavoisier called "this liquid whatever it was—an igneous liquid, matter of heat and fire", caloric (Latin *caloricum*). However, he was not fully convinced that such a liquid existed. He expresses his doubts, saying:

> Precisely speaking, we need not assume that caloric is a real matter; it suffices that it be any repulsive cause which repels the molecules of matter and allows to conceive of phenomena in an abstractive and mathematical way.
>
> Is heat a modification of caloric, or is caloric a modification of light? Provisionally, one must give different names to what causes different effects. Thus we distinguish caloric and light, not denying that they have common properties [36, Vol. I, p. 5].

Thus Lavoisier included caloric and light in the imponderable fluids. In the 18th century, the notion of such liquids was quite widespread. Electricity was then considered to be an imponderable fluid; but it was not quite clear whether only or two electrical fluids should be assumed.

In the period preceding the rise of modern chemistry in France, Polish culture in the 18th century remained in crisis. Only a few artisans practised chemistry. It was not until the reorganization of the Polish school system by the National Education Commision (the first Ministry of Education in Europe) that, in 1783, Jan Jaśkiewicz and Franciszek Scheidt began reading a course of lectures on chemistry at the Jagiellonian University, in which they propagated the ideas of Lavoisier. These ideas were accepted without strong dissent because there were not in fact any advocates of the phlogiston theory in Poland. Also, a pupil of the two Cracow lecturers, Jędrzej (Andreas) Śniadecki who lectured on chemistry at Wilno University from 1797, presented the foundations of Lavoisier's chemistry in the first Polish textbook of chemistry *Początki chemii* (The Principles of Chemistry) published in 1800. Like Lavoisier, he states that

... the terms of the elements or simple bodies (...) will signify the ultimate effects of our analyses rather than the natural beings [73].

The list of those elements agrees with that of Lavoisier's; it thus includes light and caloric. Śniadecki supplements his list in the next editions of 1807 and 1816; in the latter we find, besides light and caloric, also electric and magnetic fluids. In the second and third decades of the 19th century, the descriptions of the properties of those four 'elements', then commonly referred to as imponderable elements, frequently extended to a half of the whole space of chemical textbooks. However, Śniadecki calls them 'radiant bodies' and characterizes their properties in the following way:

Our knowledge of radiant bodies is so far very limited and mostly speculative. There are even some who deny their existence as separate bodies. However, in a specific manner they do act upon us as bodies, and as bodies they act upon other beings both animate and inanimate, which are due to them, attracted, refracted, repelled, concentrated, or rarefied, giving thereby origin in themselves to different varieties (...). For example, it is impossible to imagine the existence of animate beings without heat and light. The Sun, which is the source of both, is the prime animator of all the beings and the spirit of light. Without the Sun everything would have been coagulated and dead.

2.8. THE IMPONDERABLE ELEMENTS

Fig. 10. Jędrzej Śniadecki, acc. to A. Wrzosek, *Jędrzej Śniadecki*, Kraków, 1910.

Are therefore the radiant bodies, whose existence we have found to be indubitable, specific beings differing from one another like gases or other larger bodies? Or are they only the specific states of other known or unknown beings? For example, ice, water, and water gas are different states of the same body. That is to say, are they eventually different states and ways of manifesting the same being? It is impossible to assert upon it with certainty due to our limited knowledge and methods. In my view, according to the present state of our knowledge we can assume and distinguish four radiant bodies, namely: light, caloric, electricity, and magnetism (...).

I call them radiant bodies because when in the free state, i.e. free of any compounds, they take a rather rapid course and diffuse in the form of rays; and I hold that their state is much more volatile and rarer than that of gases. This form of denominating them is also, I believe, much more appropriate than that done by other authors (who wrote about the imponderable bodies—

R.M.), and it conveys a simple but important idea that these bodies can, by their combinations, pass to other aggregation states, including the most solid ones (...) [51, p. 34].

Next Śniadecki discusses the role of heat and caloric in chemical reactions:

It is impossible to conceive of the rarefication of bodies in any other way than by the repelling of their tiniest particles from one another; therefore, caloric must be regarded as the cause of repulsion of these particles, and thereby its action is directed against attraction and thus weakens aggregation. (...) All the known bodies are solid or liquid, or volatile, or eventually radiant. The first three bodies are generally held by Chemists and Physicists as the states of different density; therefore they believe that these bodies depend upon a different amount of caloric accumulated between their particles, i.e. they depend upon the forces of repulsion. In it, i.e. in different intensities of the action of caloric, they seek the cause of fluidity and volatility of bodies. It may be thus conjectured that every body could go through all the three states of concentration if it were given or taken as much caloric as required [51, p. 40].

Let us note that Śniadecki predicts that all substances can occur in all the three states of aggregation; thus in principle all gases can be liquefied. He states explicitly in the first edition of 'The Principles of Chemistry':

(...) if bodies are liquid or volatile and cannot be turned into solids, it should be concluded that we have not sufficient means of cooling them, and that the most intense cold that we can artificially procure is still some degree of heat in respect of absolute coldness (*frigus absolutum*) [73].

It should be realized that Śniadecki expressed this view in the time when scientists were unable to liquify even chlorine.

Like Lavoisier, Śniadecki believed that the appropriate form of all the elements was that of a solid body. He treated gases and liquids as solutions, or even combinations of those 'solids' with caloric. The state of matter depended, he believed, on the ratio of the amount of caloric to the amount of a solid. We can clearly notice here the influence of the law of multiple proportions published by Dalton only a few years earlier (cf. Chapter 3).

Concluding these considerations about caloric, Śniadecki writes about its origin:

2.8. THE IMPONDERABLE ELEMENTS

Therefore, there are two general sources of heat, i.e. the radiant caloric; firstly, condensation of bodies due either to mechanical or chemical or any other forces whatsoever; secondly, chemical combinations, and particularly those accompanied by condensation [73].

It seems that by the release of heat due to the condensation of bodies, the Polish chemist tried to account for the heat phenomena observed during mutual friction of bodies; in other words, on the basis of the substantial theory he explained the phenomena which laid the foundations of the kinetic theory of heat.

The problem of 'radiant bodies' was also discussed by Śniadecki in a separate treatise entitled *Objaśnienie niektórych punktów nauki o ciepliku* (An Explanation of Some Points Concerning the Science of Caloric) which was read at the session of Wilno University on 15 March 1815. In the treatise he rejects the commonly applied name—*principia imponderabila* (imponderable bodies) because it does not express "the nature and effect of these bodies". He also rejects the term 'ethereal bodies', derived from 'ether', to the trepidation of which Rumford attributes all the phenomena of light and heat. He describes the action of the radiant bodies in the following way:

> These entities, by their radiant power, can expand all non-radiant bodies upon which they act and which they enter, each of them in a proper degree, since their radiance is probably unequal. Hence, it follows that all these entities, having the power to expand bodies and weaken their bonds, should all in some degree favour chemical combinations and in a higher degree they should solve these combinations.
>
> Finally. The radiant being of these entities may either result from their nature and be equally specific to all of them or it may be a property of one (entity) which is found in all others and thereby they become radiant, or eventually it may depend on some specific entity unknown hitherto, which gives such a property to all of them together (...). Thus the conjecture concerning ether not supported by any experiment is unnecessary, and its nature is obscure and unknown (...). In my oral accounts I have for a long time expressed an opinion that: since the combinations of solid bodies with caloric gives them the form of liquids, and the combination of the latter with a greater amount of caloric gives rise to gases, therefore, if the latter could combine with a new, far greater amount of caloric, they should give rise to bodies of far rarer concentration than gases, i.e. the radiant bodies.
>
> In accordance with this reasoning the first and only source of radiance would

thus be in caloric, which however, must remain a mere conjecture until it is confirmed or rejected by an experiment [74].

The fragments from Śniadecki's writings quoted above show that his views concerning light and heat as chemical factors were an evident continuation of Lavoisier's views. Both scientists held caloric to be the cause of heat, "whatever it is". They both believed that caloric counteracts the attraction of molecules, and that the changes in the states of aggregation result from a differing caloric content. Śniadecki developed these ideas even further by writing about the evidently chemical combination of caloric with chemical compounds; caloric was supposed to lose then the properties which it exhibited in the free state. The Polish investigator attempted to employ quantitative chemical laws to describe the 'combinations' of caloric with chemical elements. He included light and heat in the fourth state of aggregation—rarer than the gaseous state—and gave it the name of a radiant state. He also attributed this state to electricity and magnetism. Such an approach enabled him to reject the hypothesis of the cosmic ether introduced by scientists who treated heat as the motion of corpuscles. The conception of material medium which fills space, i.e. Aristotelian ether, was necessary for them to explain the transfer of motion through the vacuum. Their approach excluded Newtonian interaction at a distance. Śniadecki opposed the ether hypothesis. We have already quoted some fragments of his texts which illustrate his position, but his most explicit statement in this respect can be found in his lecture delivered at the session of the Society of the Friends of Science in Warsaw in 1805:

(...) such a science having caloric as an imaginary body, assumes two delusive appearances: trepidation of corpuscles and ether [75].

The two major problems with which investigators engaged in the problems of light and heat were dealing: (1) the possibility of a quantitative approach to heat phenomena; (2) the transfer of light and heat through a vacuum—were solved during the course of the 19th century after the establishment of the concept of energy in physics. The adherents of the kinetic theory of heat imagined heat motion as chaotic. In the second half of the 18th century, analysis

2.8. THE IMPONDERABLE ELEMENTS

of straight-line motion led to the formation of the concept of 'live force' which was transformed by Thomas Young in 1807 to the term 'energy'. However, Young was able to define only the energy of translatory motion. Nonetheless, heat could not have been caused by translatory motion, so scientists in the first half of the 19th century failed to define quantitatively the energy of heat motion on the basis of the kinetic theory. It was James Prescott Joule who in 1845 found a quantitative way of comparing the energy of chaotic motion to that of straight-line motion. Such is the significance of comparing an increase in the temperature of a liquid in which a vane is turned by a bob falling down rectilinearly, or an increase in the temperature of a gas compressed by a rectilinear rise of a column of mercury, with the motion energy of this bob or the mercury column. On this basis, in 1847 Hermann Helmholtz formulated the first law of thermodynamics, in his work entitled *Über Erhaltung der Kraft* (On Conservation of Force). The term 'energy', which is more adequate in this case, was not yet sufficiently widespread in 1847 and that is why Helmholtz employed the term 'force'. Unfortunately, this terminological inaccuracy has lingered on in many branches of chemistry until today. For example, we speak of the force of a chemical bond but the value of that 'force' is expressed in units of energy. The problem of light and heat transfer through a vacuum was finally solved by the theory of light published in its ultimate form by James Clerk Maxwell as late as 1873.

However, an analysis of the concept of caloric, particularly in the version provided by Śniadecki and extended by him to other 'radiant beings' leads to the conclusion that these beings are nothing else but a means of a quantitative presentation of energetic transformations; the only possible way, before the final establishment of the concept of energy. If we replace the word 'caloric' in the quoted fragments of Śniadecki's writings by the term 'heat energy' and 'radiant beings' by 'energy' or 'radiant energy', we shall obtain a text which cannot be rejected by a 20th-century scientist.

How, then can we account for the 'conjecture' that all substances can be transformed to the radiant state of aggregation the conjecture that had to be supported or overthrown by experiment?

2.9. Significance of the Term 'Element' in the 19th and 20th Centuries

The way in which the concept of a simple body or an element was conceived early in the 19th century will be illustrated by a quotation from the textbook *Lehrbuch der Chemie* written by Jöns Jacob Berzelius, who was regarded as the greatest authority on chemistry during his lifetime. Berzelius classifies chemical substances in the following way:

Bodies which occur on the Earth are divided into simple, indecomposed, and composite.

(1) Simple bodies are those bodies of which we may believe with certainty that they are not composite and which occur in every respect as constituents of the remaining Nature...

(2) Indecomposed bodies are those bodies of which we may presume that they are not simple but we have not decomposed them into simpler elements; if these bodies are composite, we do not know their constituents at all.

(3) Composite bodies are those bodies which can be decomposed by chemical means into simpler constituents [76].

Berzelius made thus a distinction between indecomposed bodies and simple bodies, i.e. elements; he thus departed from the definition of the element as the actual end of analysis, although we shall come across such a definition in some textbooks published later. Moreover, like Lavoisier and Śniadecki, he had strong doubts about the nature of the imponderable elements, i.e. 'radiant beings'.

He expressed his doubts in a more explicit way than the former two investigators. However, of all the formulations made at that time with respect to the properties of the 'imponderable elements', those of the Polish investigator bear the closest analogy to the properties of energy.

Thus, the correctness of Boyle's hypothesis, formulated in 1661, that the elements are not carriers of properties but the simplest constituents of bodies was finally recognized at the turn of the 18th century. It was also realized at that time that the properties of a compound are not a sum of the properties of its constituents. The four qualities that were held by Aristotle as being basic were rejected, and other criteria of distinguishing the elements were adopted, namely:

2.9. SIGNIFICANCE OF THE TERM ELEMENT

valency and atomic masses; and, in the 20th century, the configuration of electrons on external orbitals. On the basis of these properties about 90 elements were identified. Moreover, throughout the 19th century the concept of the element was closely connected with the concept of the atom, which will be discussed in the next chapter.

The discoveries made at the turn of the 19th century shook the concept of the element based on atomic masses. Thanks to the new physical methods it was possible to break down the atoms which formed the chemical elements; they ceased to be the end of analysis. From that time elementary particles became the target of investigation. In order to distinguish them the concept of mass, supplemented by the concepts of electrical charge and nuclear spin, no longer sufficed. New elementary properties were being sought in order to systematize the set of these new particles whose number continually increased as a result of experimental discoveries. Heisenberg discovered theoretically that elementary particles of near-identical masses can have different electrical charges. This possibility is governed by a new elementary property referred to as isotopic spin; thanks to this property, one can distinguish, for example, two forms of a nucleon: a proton and a neutron. It was necessary in the subsequent years to determine some other elementary properties which differentiate elementary particles.

Today, physicists are convinced that the great number of different kinds of elementary particles (several hundred) is due to different arrangements of tinier, indistinguishable particles called 'quarks'. A separate problem is what elementary properties differentiate the particular quarks.

In this chapter we have tried to show the various modifications of the concept of the element throughout the ages. We still do not know what matter is, and we are continuously searching the Greek *prote yle* (primary matter). We endeavour to find the elementary constituents of matter, and—like Aristotle—we give a good deal of thought to the elementary properties that differentiate them. The element has ceased to be regarded as a carrier of properties because we now know that the properties of compounds and even of solutions are not a sum of the properties of their constituents; sometimes, indeed they are essentially different from that sum.

We realize that the present-day ordinary term 'element' or more precisely 'chemical element' does not denote the ultimate physical element sought since ancient times. It is even difficult to define this chemical element. Nowadays an element is frequently defined as a collection of atom with an equal number of protons. This is, however, a summary definition which includes all isotopes of equal number of protons as well as atoms of different degrees of ionization and excitation, whereas we know today that the degree of both ionization and excitation as well as the structure and composition of the nucleus of particular isotopes affect the chemical and physical properties of a cluster of atoms.

Thus, the common concept of the element is, so to speak, a conventional, 'frozen' state of this concept. As a matter of fact we still accept Boyle's definition of the element as a simple constituent which was later redefined a little more accurately by Berzelius. Subsequent modifications of the concept of the element took into account more recent scientific findings, but did not change its essential sense. However, they have never been fully satisfactory, which has been shown by the example of one of the most frequently used definitions.

CHAPTER 3

The Elementary Particle of Matter

3.1. The Conceptions of Continuity and Discontinuity of Matter in Antiquity

The problem of the heterogeneity of the basic constituents of particular substances was being solved, as we have shown in the preceding chapter, by seeking the simplest bodies, i.e. elements. However, this did not settle the problem of the structure itself, or more precisely, the texture of matter. Two parallel views on this problem existed in ancient times. According to one of them, matter was held to be continuous and, according to the other, discontinuous or discrete. The latter conception is usually associated with the name of Democritus, whose works have been preserved only in small fragments. We know the views of that philosopher from accounts of his adherents, from a didactic poem *De rerum natura* written by the Roman poet Lucretius in the 1st century BC [24], and above all from the works of Aristotle. Aristotle was, however, critical about the conception of discontinuous matter, and, on the contrary, he went on to argue that it must be continuous.

In one of his first works, 'Physics', Aristotle, having considered a possible number of different elements, proves that

If according to Leucippos and Democritus, principles are numerically unlimited, then as Democritus had assumed, they are homogeneous, and only differ in shape and form from one another, or they are of various kinds [23, 184b].

Similarly in another work, 'On Generation and Corruption', he wrote that

Democritus and Leucippus say that there are indivisible bodies, infinite both in number and in the variety of their shapes, of which everything else is composed [23, 314a].

These indivisible particles were called 'atoms' by Democritus (from Greek *atomos*—indivisible). Atoms were believed to have different shapes and "the generation and corruption of bodies were attributed to their separation and union, whereas transmutation was due to the arrangement and position of their constituents" [23, 315b]. The form of a body and its appearance seemed to be dependent on the arrangement and position of the atoms of its substance. Each substance, according to Democritus, had its own atoms (likewise we consider today that each substance consists of its own molecules; however, they are chemically divisible, contrary to Democritus' atoms). Therefore, there must be as many kinds of atoms as there are kinds of substances, that is practically infinitely many. As we have seen in the preceding chapter, Aristotle rejected the feasibility of the existence of an infinite number of elements because he argued nature would then have been unknowable.

We have mentioned in Chapter 1 that a characteristic feature of the ancient outlook was a juxtaposition of contrarieties. According to this approach the contrary to atoms as being was non-being or vacuum. For that reason the atomists believed that atoms existed in a vacuum. According to Aristotle:

(...) void is said to be that in which the presence of the body, though not actual, is possible [23, 279a].

Since, in accordance with Aristotle's argument contained in 'Physics', motion can be effected only in a material medium, no motion would be feasible in a vacuum, hence the world would be static and unchangeable. However, we observe in nature a lot of change and different kinds of motion, therefore, Aristotle concluded, there is no vacuum in the Universe.

Let us note that Aristotle's reasoning is not so absurd as it might have seemed. Aristotle's vacuum is a homogeneous space devoid of matter. It is a geometrical space which does not exist in nature. At every point of physical space there act different kinds of force fields which are a form of matter. Points through which radiation penetrates are not void either, because there is an electromagnetic field in them.

Apart from the above two objections, Aristotle raised a few others

3.1. CONTINUITY AND DISCONTINUITY

against the atomists. We have mentioned in Chapter 1 the four kinds of determinants by which Aristotle explained all phenomena. In the atomistic picture of the world there was no place for a final cause; the atomists were unable to show the purpose of phenomena. Moreover, they remained in a vicious circle when they attempted to explain the shapes of atoms by the properties of a substance which was composed of them, and—conversely—the properties of a substance by the shapes of atoms. For example, some atomists explained the 'piercing' sensation (burning), caused by fire, by the tetrahedral shape of the atoms of fire, since a regular tetrahedron has the sharpest verticles of all regular solids, hence it is the most 'prickly' one.

Although the atomists descended mostly from the Eleatics, many of them recognized, just like the Ionian philosophers, the existence of simple bodies. These bodies, regarded as perfect, must have had, in Plato's opinion, perfect shapes as well, and their atoms must have been in the shape of regular polygons, which have been referred to until today as the Platonic solids. Thus the atoms of the Earth were believed to have the shape of a regular hexahedron or a trigonal bipyramid; the atoms of water had the shape of an icosahedron—a figure which resembles a sphere; and the atoms of air are mobile due to their having the shape of an octahedron. Tetrahedral atoms were attributed, as we have already said, to fire, whereas the atoms of the heavenly matter (ether) were believed to be dodecahedra whose faces were regular pentagons because a star inscribed in such a pentagon has particular features of geometrical harmony, and therefore it was also a symbolic sign of the Pythagoreans. Atoms of compound bodies were believed to have more complex shapes, not so ideal as the former, but their shapes—according to the atomists—also predetermined the properties of substances. For example, the salty taste of salt was believed to be due to the serrated shape of its atoms.

The peripatetics also objected that by means of Democritean atoms it is impossible to explain human emotions, particularly love and its opposite—hatred. Besides, the atomic theory could *ex post facto* account for every observation; however, it was not possible to predict any phenomena.

The peripatetics' critique prompted some followers of Democritus to modify his original theory. The Greek philosopher Epicurus, who lived in the 2nd century BC held that:

(...) atoms being indivisible and full bodies of which compounds are made and into which they decompose (...) have a countless number of forms. In each form alike atoms are in an infinite number, and the multiplicity of their forms is not infinite but innumerable [77].

Similar opinions were expressed by other ancient commentators on Aristotle, such as Alexander of Aphrodisias (2/3rd c.), and Simplikios of Cilice (6th c.) [78].

Aristotle and the peripatetics contested the atomic conception by the conception of continuous matter. In the treatise 'On Heaven', Aristotle writes:

(...) a continuum is that which is divisible into parts always capable of subdivision, and a body is that which is every way divisible [23, 268a].

and in the conclusion of this treatise:

Some continua are easily divided and others less easily (...) and air is more so than water, water than earth. Further, the smaller the quantity in each kind, the more easily it is divided and disrupted [23, 313b].

Since Aristotle does not recognize the existence of the vacuum, he believes that all the parts into which a body is divided must closely adhere to one another. However, it follows from the 'Physics' that all these parts need not be identical. Aristotle believes that bones are part of an animal, and fruit part of a plant. In his treatise 'On Generation and Corruption' he distinguishes mixtures which are different, he says, from compound or homogeneous bodies. The former consist of perceptible parts whereas the latter do not. The latter bodies include, of course, according to Aristotle, not only substances which are now called chemical compounds, but solutions, amalgams, and alloys as well.

According to the Peripatetics, the properties of a body were a sum of the properties of its constituents; it thus seemed that one could change the properties of a given body by adding a certain amount of a substance abounding in an element which carried the desired property. In this way, they believed, one could obtain a substance which would have desired properties, too. This reasoning favoured attempts to obtain new materials; thus—in spite of certain misconceptions—it was much fruitful than that of the atomists. Undoubtedly, this was one of the reasons why the ideas of the peripatetics dominated practically throughout the whole period of Greek science.

The other significant cause of the greater popularity of the peripatetic ideas than that of the atomistic ideas was the fact that the latter did not include the sphere of human sensations. Ancient and mediaeval philosophers, as has been stressed in Chapter 1, attached a great importance to the uniform view of natural phenomena. Thus, failure of the atomistic theories to deal with the animate world and human sensations was the second factor which prompted the ancient philosophers to reject them.

3.2. Rebirth of the Corpuscular Theory in the 14th–18th Centuries

Christian thought in the first millennium was based in great measure on Plato's philosophy, and in the later Middle Ages, as we have mentioned earlier, the views of Aristotle were generally adopted; however, the ideas of Democritus were also studied. At first they did not find a wide response, although some works containing atomistic views had been translated from Arabic into Latin by Adelhard of Bath as early as the 12th century.

The peripatetic views were still quite sufficient for the qualitative description of phenomena. However, in the quantitative description, it was easier to apply the corpuscular approach, and therefore, with the development of quantitative methods this approach became widespread. In the mid-14th century this conception was propounded by the Paris scholar Nicholas of Autrecourt, and in the following century, by the outstanding German philosopher Nicholas of Cusa. The Italian philosopher Giordano Bruno, who was burnt at stake in 1600 by the Inquisition, also propagated corpuscular views.

In the 17th century, quantitative research began to prevail in physics over qualitative methods, and at that time the corpuscular conceptions met with a wider acceptance. The French philosopher Pierre Gassendi recalled the views of Epicurus. According to Gassendi, the atoms of particular substances have different masses; and in order to emphasize that the atom is a particle of a very small mass, the philosopher coined a Latin term *molecula* which is a diminutive of the word *moles* (mass). The name *molecula* was still used interchangeably with the name atom as late as the early 19th century. Throughout the 17th century many physicists, iatrochemists and physicians employed corpuscular concepts in their writings, e.g. the Italian physician Giovanni A. Borelli accounted for the phenomenon of capillarity by the interaction of liquid molecules with the walls of vessels. The German philosopher Joachim Jungius believed that even "Fire consists of the atoms of *halitis pingui* (burning breath) and perhaps of the atoms of flame" as well. He also held that "nothing is harder than the atom" [50].

Robert Boyle was a dedicated adherent of the corpuscular texture of matter. In his work 'The Sceptical Chymist' he wrote:

It seems not absurd to conceive that at the first production of mixt bodies, the universal matter whereof they among other parts of the universe consisted, was actually divided into little particles of several sizes and shapes variously moved [32, p. 30].

Boyle points out further that a proof for the existence of such particles is the rise, decay, nourishment and destruction of bodies, as well as the fact that these particles, being very small, can sometimes be seen under a microscope or one can prove their existence by decomposing complex bodies or by subjecting them to the action of 'spagyric' fire. Boyle attributes the original conception of the existence of such small particles to Epicurus or Moses, since the former proved that the world is made up to atoms moving in a vacuum under the influence of an inner cause, while the latter is believed by Boyle to have said that

(...) the great and wise Author of things did not immediately create plants, beasts, birds, etc., but produced them out of those portions of the proexistent, though created, matter, that he calls water and earth, allows us to conceive that the constituent particles whereof these new concretes were to consist, were variously moved in order to their being connected into the bodies they were, by their various coalitions and textures, to compose.

Neither is it possible that these minute particles diverse of the smallest and

3.2. REBIRTH OF THE CORPUSCULAR THEORY

neighbouring ones were here and there associated into minute masses or clusters, and did by their coalitions constitute great store of such little primary concretions or masses as were not easily dissipable into such particles as composed them [32, p. 30].

Next, Boyle provides the following argumentation:

I consider that it very often happens that the small parts of bodies cohere together but by immediate contact and rest, and (...). I will not peremptorily deny, but that there may be some clusters of particles, wherein the particles are so minute, and the coherence so strict, or both, that when bodies of differing denominations, and consisting of such durable clusters, happen to be mingled, though the compound body made up of them may be very differing from either of the ingredients, yet each of the little masses or clusters may so retain its own nature, as to be again separable, such as it was before [32, p. 87].

It follows from his further statements how he conceived of the corpuscular texture of matter:

(...) is it be granted rational to suppose, as I then did, that the elements consisted at first of certain small and primary coalitions of the minute particles of matter into corpuscles very numerous, and very like each other, it will not be absurd to conceive, that such primary clusters may be of far more sorts than three or five [32, p. 96].

As can be seen, Boyle strongly favoured the view that matter has a corpuscular, i.e. discontinuous texture. He believed that matter is made up of tiny lumps and clusters which, in turn, make larger combinations. This idea was reformulated more precisely by Newton half a century later. His corpuscular views, as can be shown by the following quotations from the second edition of his work entitled 'Opticks' (1717), were based on theological and teleological assumptions:

All bodies seem to be composed of hard particles (...). It seems probable to me that God in the beginning formed matter in solid, massy, hard, impenetrable, movable particles, of such sizes and figures, and with such other properties, and in such proportion to space as most conduced to the end for which he formed them; and that these primitive particles being solids, are incomparably harder than any porous bodies compounded of them; even so very hard as never to wear or break in pieces; no ordinary power being able to divide what God Himself made One, in the first creation. While the particles continue entire they may compose bodies of one and the same nature and texture in all ages; but should they wear away or break in pieces, the nature of things depending on them would be changed. Water and earth, composed of old worn particles, and fragments of particles, would not be of the same nature and texture now, with water and earth composed of entire particles in the beginning. And therefore that nature may be lasting, the changes of corporeal things are to be placed only in the various separations and new associations, and motions of these permanent particles; compound bodies being apt to break, not in the midst of solid particles, but where these particles are led together, and only touch in few points [79].

Ninety-three years later, John Dalton would transcribe the whole of the above fragment in his 'Diary' [80, p. 123].

How Newton conceived of the texture of matter in more detail is described in the work quoted a few pages earlier:

Now, the smallest particles of matter may cohere by the strongest attractions, and compose bigger particles of weaker virtue; and many of these may cohere and compose bigger particles whose virtue is still weaker, and so on for divers successions until the progression ends in the biggest particles in which the operations in chemistry, and colours of natural bodies depend, and which by cohering compose bodies of a sensible magnitude [79].

An interesting conception of the atomic texture of matter was put forward in the mid-18th century by the Croatian scientist Rudjer Bosković. He held that in continuous matter there exist points which are centres of vortical motion. These points were believed to play the role of atoms; vortical motion, which they generated, caused other points to tend to them, the result of which was, as he believed, the effect of attraction. Bosković's theory was not understood by contemporary investigators and did not exert any influence on the development of science. On the other hand, the corpuscular approach to the texture of matter became widespread throughout the 18th century. We shall illustrate it by quotations from the works of the Russian scientist Mikhail Lomonosov, written in the years 1739–1750, and by fragments of one of the first Polish textbooks of physics written by Father Józef Rogaliński and published in Poznań in 1765 under the title *Doświadczenia skutków rzeczy pod zmysły podpadających* (Experiments of the Efffects of Things Submitted to the Senses) [34].

The quotation that follow should be treated only as an illustration of the sort of reasoning which was quite common in the 18th century.

3.2. REBIRTH OF THE CORPUSCULAR THEORY

Only a few works of Lomonosov were published in his lifetime in Latin by the Petersburg Academy of Sciences. The majority of them remained in manuscript and, having been published only in the 20th century, they did not have any influence on the development of science. Rogaliński's textbook was only of significance for the advancement of Polish science.

In his early dissertations, written by Lomonosov when he was still a student (*Specimen physicum de transmutatione corporis solidi in fluidum* from 1738 and *Disertatio physica de corporum mixtorum differentia* from 1739), the causes of different states of aggregation of bodies are explained by a different degree of binding of particles which make up these bodies. This view is illustrated by diagrams which show the mutual arrangement of particles. The above-mentioned dissertations of Lomonosov were published for the first time in 1934. In 1743, the Russian scientist wrote a treatise entitled *De particulis physicis insensibilius corpora naturalia constentituentibus* in which he emphasized that "particular physical particles are impenetrable, they govern the inertia force, motion, and peace". Among those particles there also exist "indivisible particles, i.e. ones which cannot be divided into smaller particles in a natural way" [35, Vol. I, p. 279]. Also, in a dissertation on saltpetre written in 1749, the interacting particles are represented in the form of spheres of different diameters and different surface roughness. This work was published in Novi Commentari Acadaemiae Scientiarum Petropolitanae. In another dissertation, published at the same time and entitled *Tentamen theoriae de vis aeris elastica*, Lomonosov does not doubt that the elastic force of air "comes from interaction of its atoms". Further, he states that "it is necessary that the atoms of air which directly interact be in mutual contact". And the possibility of compression and expansion is explained by the fact that the degree of interaction of atoms of air depends on the degree of heat; heat, in turn, depends in his view on the "rotational movement of the particles of a hot body" [35, Vol. II, p. 105].

As can be seen from the above quotations, the corpuscular view of the texture of matter was for Lomonosov a starting point to explain many phenomena.

The work of Rogaliński was designed as a textbook; therefore we find there a survey of different views of the texture of matter. The author writes:

> First Democritus and Epicurus, the Greek Sages, brought up in pagan blindness, said that the primary matter is nothing else but fine, inseparable dusts which can be neither seen nor perceived by the senses. Those men, who did not know God, said that: these dusts are eternal, uncreated, infinite, in number and thus by flying, revolving, and turning in a confusion for centuries, they have been arranged quite haphazardly so that they have made up the Heaven, stars, earth, water, and the whole visible world (...). This pagan and irrational science which attributed God's property, i.e. eternity, to the base dusts was mocked by the following centuries, but Gassendi, the famed Frenchman in the last century, corrected this impiety by teaching that these dusts are not everlasting but are created by God; however, they can be neither born nor destroyed; they are composed of different particles, but they are so coherent that no natural force can divide these particles; for upon creation God cemented them in such a way that no force can tear them away from one another [34, p. 4].

The last fragment of the above quotation proves that Rogaliński considered Democritean atoms to be complex forms and not the simplest structural particles. In accordance with the views prevailing in the 18th century, that the element is the limit of analysis, Rogaliński also treats the atoms as an end of the possibility of the division of bodies. It is important at this point to realize what Rogaliński as well as other authors understood by the term 'divisibility'. He discusses the problem of divisibility of bodies in his textbook. He observed some substances under a microscope.

> All major bodies, which we can perceive with our eyes or touch them, can be divided into many particles which will be the smaller the more we divide them; eventually they will be so fine that we shall lose sight of them. So, for example a locksmith's file turns a large piece of iron into dust (...). However tiny these dusts might seem to us, they can still be divided into tinier particles by different means. Different crafts show us that these same dusts which we can hardly see with our eyes because of their tininess are in fact like some minute mounds consisting of even tinier dusts (...). A piece of wood thrown into the fire will soon cease to be wood because not only these tiny mounds which make up the whole are separated, but even the particles which constituted the very mounds are disintegrated by fire and so scattered they make up smoke, fire, and ash. Eventually, the same particles, commonly unlike to one another (whose union, however, made various alike mounds) cannot be

3.2. REBIRTH OF THE CORPUSCULAR THEORY

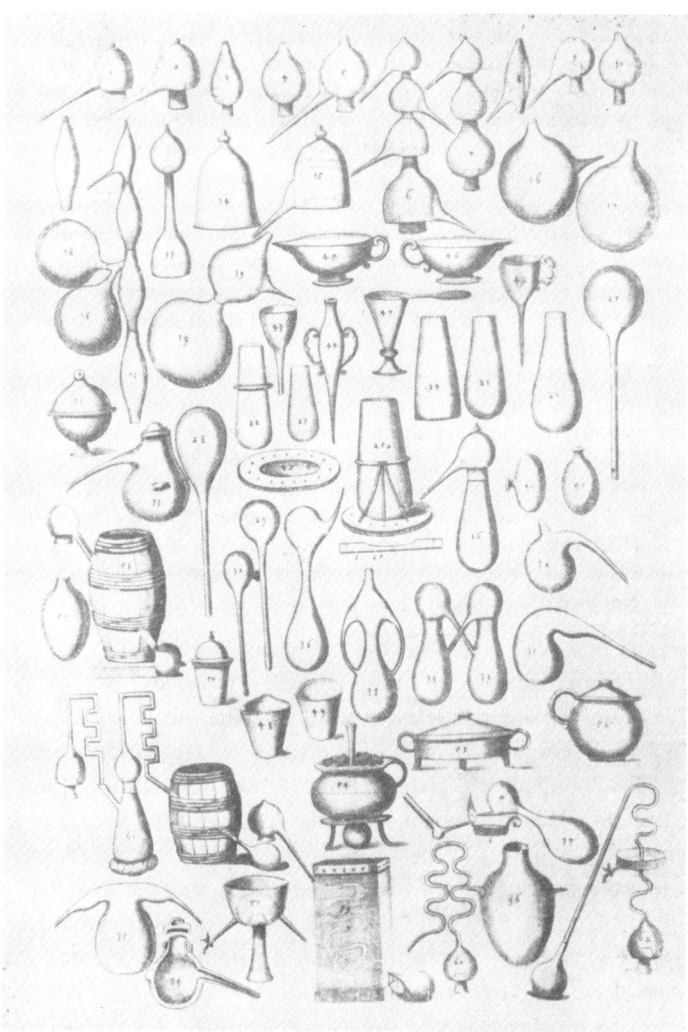

Fig. 11. 17th-century alchemical instruments, from *Theatro Farmaceutico*, 1682, acc. to [7].

inseparable, but again they consist of other particles. For we do not know about any entity in the world which would not consist of entities of a different kind (...). Thus when separating one thing we have reached such tiny particles that we have no way of separating them further, it must not be understood that we have reached the ultimate particle, because this same particle has smaller particles of which it was made up; however, the impossibility of division and separation of these particles is not due to the fact that they were not actually there but to the fact that we have not such tiny instruments for separating them, or that because of their baseness we can neither capture nor see them (...).

Therefore, those who teach that bodies are infinitely divisible speak well, and those who teach that they are not infinitely divisible also speak well, but in a different sense. I shall explain it: if we speak about separation only, as idea or concept. I shall willingly agree that in this reasoning the tiniest dust is infinitely divisible (...). If, however, we speak about actual separation? It is of course evident that bodies are not infinitely divisible because we do not know whether these particles which are in mutual contact can be actually separated from one another? (...) No experiment has confirmed this to us so far. And that in natural things the most obvious evidence of all truths should be derived from experiment, we can thus infer that so far we know nothing for sure about (...) the finite or infinite separation of bodies (...). For even if God the Creator had created infinitely divisible particles, he could have preserved His work by other means, e.g. by leaving behind in the world only weak and insufficient ways of separating these particles which in fact (if God would have allowed) could have been separated [34, p. 2].

As can be seen, like Newton, Rogaliński believed that matter gradually forms more and more complex particles, but every consecutive 'complex' of particles is weaker than the previous one. The Polish author emphasizes that the degree of separation depends on the applied method of division. What various authors meant by separation and the problem of methods of separation will be discussed further on.

While analysing texts from the 17th century and the first half of the 18th century, let us also note that scientists seemed to have had only a vague idea of the difference between an homogeneous mixture and a chemical compound. In Chapter 2 this has been shown by a quotation from Rey's work, published in 1630. Rogaliński did not notice this difference either. It finds evidence in the descriptions and interpretations of the following experiments. Rogaliński proposes dissolving a few pieces of copper in *aqua fortis* (nitric acid) and observing them until they disappear, and the solution turns green. He believes that this is the effect of separation of the 'mounds' of copper by the 'mounds' of nitric acid which act as 'sharp axes or knives on wood'.

3.2. REBIRTH OF THE CORPUSCULAR THEORY

He says that the particles of water act in a similar way on the particles of sugar when it is dissolved. He notices, however, that when *spiritus vini* (alcohol) is added to the water solution of sugar, "water goes to wine and abandons sugar", the latter being precipitated on the bottom of the vessel. Rogaliński holds that an analogous process takes place when a piece of iron is immersed in a previously prepared solution of copper in nitric acid; iron is then covered with copper.

It appears that for Rogaliński the mechanism of precipitation of sugar from a water solution saturated by addition of rectified ethyl alcohol was the same as that of percipitation of copper from the water solution of cupric nitrate into which an iron rod was immersed.

Also Lavoisier, who in principle distinguished chemical compounds from mixtures, did not wholly understand this difference. In his work *Traité élémentaire de chimie* published in 1789 he states:

(...) atmospheric air is composed of two elastic fluids of different and so to say opposite nature [36, Vol. I, p. 94].

and further he writes:

(...) water is not a simple elementary substance, but is composed of two elements, oxygen and hydrogen; which elements, when existing separately, have so strong an affinity for caloric as only to subsist under the form of gas in the common temperature and pressure of our atmosphere.

As can be seen, Lavoisier used the same term 'compose' to denote both the mixing of oxygen and nitrogen in the air and binding of oxygen and hydrogen in water. Besides, as we have already mentioned, both Lavoisier and Śniadecki held that all elements are solid bodies and only by binding with the caloric could they assume a liquid or gaseous state. Lavoisier also employed the same term 'combinaison' to denote both the binding of hydrogen with caloric and the binding of carbon with oxygen in the description of the same experiment on the decomposition of water cited in the previous chapter. He used the same term to denote the binding of metals in metal alloys as well as the compounds of metallic and non-metallic substances with oxygen of different degrees of oxidation [36, Vol. I, p. 116].

This difficulty in distinguishing solutions and compounds by the end of the 18th century resulted from the lack of criteria. Such criteria were, however, soon formulated on the basis of quantitative chemical laws.

3.3. The First Quantitative Laws in Chemistry

A systematic application of the weight methods and, from the mid-18th century, of the wet methods of analysis led gradually in the 19th century to the establishment of the concept of the molecule and later of the atom. At the turn of the 18th century, many investigators, e.g. Lavoisier, or Śniadecki accepted without discussion the corpuscular texture of matter. The Democritean term 'atom' and the term 'molecule' introduced by Gassendi were synonymous in that period. Lavoisier and later Avogadro almost exclusively employed the term 'molecule'. Dalton, however, used the term atom 'whose' significance will be discussed in detail later. For the time being it will be worthwhile mentioning that in his translation of a fragment of Lavoisier's textbook Dalton replaced the term molecule used in the French original by the term atom [81]. Simultaneously, the term 'atom' was used interchangeably with the term 'element'.

The terms 'molecule', 'atoms' or 'element' had only a qualitative meaning at that time. The first quantitative concept referred to 'chemical equivalent'. The equivalence of different acids with respect to definite bases was already known in the 18th century. The term 'chemical equivalent' had already been used in 1767 by H. Cavendish to denote a certain qualitative property. In a work entitled 'Stoichiometry' (published in 1792) Jeremias Benjamin Richter made use of equivalent quantities, although he never used this name explicitly. This is how he justifies the title of his work:

Since the mathematical part of chemistry is mostly concerned with indecomposable bodies or elements, and teaches us how to determine the quantitative relations between them, I could not invent for this discipline of science a more concise and apt name than the word *Stöchymetrie* from *stocheion*, which means in Greek speech something that cannot be further decomposed, and meterein which means finding quantitative relations (p. XXIX).

This reasoning is based on the following observation: two neutral salts

3.3. THE FIRST QUANTITATIVE LAWS

which decompose one another form neutral combinations anew. The direct conclusion which I have derived therefrom could not have been other than the one (which states) that definite quantitative relations must exist between the constituent parts of neutral salts [82, p. 3].

Then assuming the value 1000 for the relative quantity of sulphuric acid (i.e. SO_2), Richter determined the values of those relative quantities of other constituent parts of compounds with an accuracy of one unit. It was, in fact, the first extensive table of chemical equivalents of the elements and compounds in history, or—more precisely—of the constituents of the compounds. The very term 'equivalent' was introduced 22 years later by W. H. Wollaston in a work entitled 'A Synoptic Scale of Chemical Elements'. Wollaston's considerations are based on Richter's reasoning and on the new concepts which were developed in the period of those twenty years. The first of these concepts included the law of constant composition. In 1799, Louis Joseph Proust stated: "It must be concluded that nature acts deep inside the Earth in no other way than it does on its surface or in the hands of man. These always unchangeable proportions are the constant features which characterize the true relations created by the arts or nature (...). Shouldn't we therefore, according to these considerations, think that natural copper carbonate will never differ from the one which differ from the one which will be imitated by the arts?" [83]. Proust asserted thus that the weight ratio of copper, oxygen, and carbon in copper carbonate always equals 5.3:4:1. Elżbieta Petruska-Madej has shown that the 'law of constant composition' had been used practically by scientists before it was finally formulated and proved by the French chemist [84].

Further in his work, on the basis of the law of the constant composition of a chemical compound. Proust attempted to distinguish between a chemical compound and a solution:

But what difference (...) one may say do you recognize between your chemical combinations and these assemblages of combinations which nature does not bind fast according to you in any fixed proportions? Is it that the power which makes metal dissolve in sulphur is other than that which makes one metallic sulphide dissolve another. I will not hasten to respond to this question, sound as it is, for fear of wandering into a region which the science of facts has perhaps not sufficiently clarified, but one will none the less conceive my distinction, I hope, when I ask, is the attraction which makes sugar dissolve in water the same or not the same as that which makes a determinate quantity of carbon and hydrogen dissolve in another quantity of oxygen to form the sugar of our plants? (...) Thus the solution of nitre (saltpetre) in water is, for me, not at all like that of azote in oxygen which produces nitric acid or that of nitric acid in potash which produces saltpetre. The solution of ammonia in water is to my eyes not at all like that of hydrogen in azote which produces ammonia [85].

Proust held molecule attraction to be the cause of the formation of both solutions and compounds, and although in both cases he used the same term 'solution', he realized an essential difference in the properties of both kinds of 'solutions'.

On the other hand in his *Essai de statique chimique* published four years later, Claude Louis Berthollet analysed more thoroughly both the cases when substances are mixed in constant ratios and when they are mixed in arbitrary ratios. On the basis of the quantitative determination of oxygen and lead content in the lead oxides he investigated, he discovered that elements can combine in variable quantitative ratios, which opposed Proust's idea expressed in his law of constant ratios:

I have shown that the proportions of oxygen in oxides can gradually change from the condition under which a combination becomes feasible to that under which we attain the final grade of saturation and if this effect does not follow it is simply because the conditions to which I have pointed are an obstacle in gradual action [86, Vol. II, p. 370].

Berthollet mentioned the following factors which might be an obstacle in gradual action: insolubility, cohesion, elasticity, and the formation of crystals. And as examples of bodies in which continuous action was observed, the French scientist gave glass, alloys, and solutions. Of course, in the case of the analysis of lead oxides Berthollet had to do with different mixes of oxides which he was unable to separate and distinguish.

Due to the above considerations, Berthollet began to realize better than his predecessors the difference between a chemical compound and a liquid or solid solution:

Chemists, astounded by the fact that in many combinations definite proportions (of elements—R.M.) were found, frequently took it as a general rule that they were formed in definite proportions, so what when, in their opinion, neutral salt receives an excess of acid or base, the homogeneous substance which is formed thereof is a solution of neutral salt in a free amount of acid or base.

This hypothesis is based on a distinction between a solution and a chemical compound which includes the properties that provoke separation and affinity produced by the compound, but it is necessary to recognize the circumstances which can cause separation of compounds in a given state, and which thus hamper the effects of the general law of affinity.

The existence of some crystalline salts, referred to for many years as acid salts, i.e. such salts in which a complete neutralization of acid has not taken place, was evidence for Berthollet of a certain inadequacy of the definition of a chemical compound. This follows from his further considerations.

In consequence of what has already been presented, saturation of two kinds must be distinguished: one being a limit of chemical action which can be exerted by one substance upon another under given circumstances e.g. we say that water is saturated with salt when it cannot dissolve more salt although the properties of salt have not undergone saturation; the other being a situation in which the opposite properties of one substance are masked by the properties of the other substance and are in equilibrium which causes (...) what we call the state of neutralization; the other saturation occurs rarely under the same conditions as the first one [86, Vol. I, p. 62].

As can be seen, Berthollet considered that differences occurred between a mixture and a compound in both liquid and solid solutions. Simultaneously, Dalton realized the principal differences between a combination of two gaseous substances and a mixture of gases.

3.4. Corpuscular Views of Dalton

We shall now examine in more detail the views of John Dalton because his considerations exerted a significant influence on the development of chemical concepts.

In his early life Dalton made meteorological observations which prompted him to speculate on the structure of air. It follows from his first textbook 'A New System of Chemical Philosophy', published in 1808, that he initially regarded air as a chemical compound of water, oxygen, and nitrogen. However, even before the year 1801, he came to a conclusion that air must be a mixture of the above-mentioned gases, which should remain separated on account of their different specific gravities, but which are brought into contact with one another as a result of slow diffusion. Like Lavoisier and Śniadecki, Dalton held that in the gaseous state.

(...) the bulk of the particle signifies the bulk of the supposed impenetrable nucleus, together with that of its surrounding repulsive atmosphere of heat [87, p. 20].

Further, Dalton wrote:

At the time I formed the theory of mixed gases, I had a confused idea, as many have, I suppose, at this time, that the particles of elastic fluids are all of the same size: that a given volume of oxygenous gas contains just as many particles as the same volume of hydrogenous [87, p. 11].

Unfortunately, the observations of reactions between gases, e.g. the formation of nitric oxide from oxygen and nitrogen, prompted Dalton to reject what we now know to be a correct statement, and led him to the false conclusion that although the particles of a given gas always have the same volume, its value differs for particular gases. This conclusion enabled him to assert further that in the case of contact of the particles of the same gas, the forces of attraction of nuclei and the forces of repulsion of their 'heat atmospheres' are equalized; however, when particles of different gases are in contact the resultant force causes both gases to be mixed as a result of diffusion.

As can be seen from the foregoing, it is essential to determine how Dalton conceived of these 'impenetrable nuclei'. This problem will be considered further.

After completing his investigation of the atmosphere, Dalton carried out measurements of the solubility of gases in liquids: he presented the results obtained and discussed them during a lecture in the Manchester Literary and Philosophical Society on 21 October 1803, and published them in an article in *Manchester Memoirs* entitled 'On Absorption of Gases by Water and Other Liquids'. In the course of his investigation Dalton increased the pressure of gases and measured the value by which this pressure decreased after a unit of time. He also noticed that a certain amount of gas was spontaneously given off from

3.4. CORPUSCULAR VIEWS OF DALTON

the solution when he decreased the pressure of the gas. From these observations Dalton drew the conclusion that gases mix with water mechanically and not chemically; and he wondered why under the same conditions the amounts of different gases dissolved depend on their kind. He wrote:

Why does water not admit its bulk of every kind of gas alike? This question I have duly considered, and although I am not yet able to satisfy myself completely, I am nearly persuaded that the circumstance depends upon the weight and number of the ultimate particles of several gases; those whose particles are lightest being less absorbable, and the others more, according as they increase in weight and complexity. (Subsequent experience renders this conjecture less probable). An inquiry into the relative weights of the ultimate particles of bodies is a subject, as far as I know, entirely new [88].

Fig. 12. John Dalton, acc. to Millington J. P., 'John Dalton', J. M. Dent and Co., London 1906.

3. THE ELEMENTARY PARTICLE OF MATTER

The basis of the conception of relative weights was formulated by Dalton in the first volume of his work:

> Whether the ultimate particles of a body, such as water, are all alike, that is, of the same figure, weight, etc., is a question of some importance. From what is known, we have no reason to apprehend a diversity in these particulars: if it does exist in water, (...) some particles of water (should be) heavier than others (...), a circumstance not known. Similar observations may be made on other substances. Therefore, we may conclude that the ultimate particles of all homogeneous bodies are perfectly alike in weight, figure, etc. In other words, every particles of hydrogen is like every particle of hydrogen, etc. [87, p. 141].

At the end of the work on absorption of gases Dalton presented a table of relative weights of ultimate particles. Although this table was published as early as 1805, it was not in fact the first but the fourth version of Dalton. We know about it from his laboratory diary published in 1896 which comprised the years 1802–1808 [80, p. 26], from his textbook, and lectures delivered at the Royal Institution in London in 1810. Of the greatest importance are six pages of notes (pages 244–249) of his laboratory diary written on his 37th birthday on 6 September 1803. It may be demonstrated that all further major publications of Dalton are in fact merely an extension of the ideas contained on these six pages.

Thus, on one page, p. 245, of the 'Diary', entitled 'Observations on the Ultimate Particles of Bodies and Their Combinations', we find the following description of the impenetrable nuclei of gaseous particles:

> The ultimate atoms of bodies are those particles which in the gaseous state are surrounded by heat; or they are the centers or nuclei of the several small elastic globular particles [80, p. 27].

On page 247 of his 'Diary', basing himself on the work of other investigators, Dalton defines how many times an ultimate particle of oxygen is heavier than that of hydrogen (analysis of water), how many times an ultimate particle of nitrogen is heavier than that of hydrogen (analysis of ammonia), and he also provides the ratio of the ultimate particles of oxygen to those of nitrous oxide. He assumes that water, ammonia, and nitrous oxide contain only one ultimate particle of each element, respectively. The 'atomic weights' determined in this way were in fact chemical equivalents. On the same page, based on the results

3.4. CORPUSCULAR VIEWS OF DALTON

of analyses of sulphur oxides performed by the French chemists, Dalton attempts to determine the ratios of the weights of the ultimate particles of sulphur with respect to oxygen and the proportions of these elements in sulphurous and sulphuric oxides. It follows from the calculations presented on this page of his notebook that sulphuric oxide should contain twice as much oxygen as sulphurous oxide.

Using such premises, on page 248 Dalton lists the weights of the ultimate particles of oxygen, nitrogen, carbon, sulphur, water, ammonia, three nitrogen oxides, both sulphur oxides, and two carbon oxides. The weights of the ultimate particles (atoms) of compounds constitute a sum of the weights of the ultimate particles of the constituent elements. On the next page Dalton presents the composition of the elements listed previously, using the circle symbols of atoms given on page 244 of his Diary.

According to his assumptions Dalton provides on these two pages the weight of an ultimate particle of water (6.66) as a sum of the weight of the ultimate particle of oxygen (5.66) and of hydrogen (1.00); the weight of an ultimate particle of ammonia being 5, given as the sum of the weight of an atom of nitrogen (4) and an atom of hydrogen (1). An atom of sulphurous acid consists, according to Dalton, of one atom of sulphur and one atom of oxygen, and an atom of sulphuric acid consists of one atom of sulphur and two atoms of oxygen. Dalton also distinguished three nitric oxides; 'nitrous gas'—a combination of one oxygen atom and one nitrogen atom; 'nitrous oxide' containing one atom of oxygen and two atoms of nitrogen; and 'nitrous acid' consisting of two oxygen atoms and one nitrogen atom. Later Dalton drew on the idea of 'nitric acid' with three oxygen atoms and two nitrogen atoms. However, he held that the ultimate particle of ammonium nitrate is a combination of one ultimate particle of 'nitrous acid', one ultimate particle of ammonia, and one ultimate particle of water.

The table of atomic weights (the weights of ultimate particles) presented on page 248 of the 'Diary' is thus the first table compiled by Dalton, and moreover the first table of this kind in the history of chemistry. It was based exclusively on the data published earlier by other investigators, hence the ratio of the weight of an oxygen atom to that of a hydrogen atom (5,66) is the result of Lavoisier's investigation into the decomposition of water which was discussed in the previous chapter.

Dalton several times modified and extended his table of the relative weights of ultimate particles (atoms) taking into account both the results of his numerous analyses of the thermal or electrical decomposition of compounds and the results obtained by other investigators. As early as a few weeks after he made the note of 6 September we find in his Diary the number 5.5 as the value of the relative weight of an oxygen atom because Dalton found that the specific weight of oxygen is 11 times greater than that of hydrogen. Together with this value is a mention:

... but 1 measure oxy(gen) seems to combine with 2 hydrogen [80, p. 84].

Whereas, in 1808, Dalton states in his handbook

An atom of steam composed of 1 oxygen and 1 hydrogen retained in physical contact by strong affinity and supposed to be surrounded by the atmosphere of heat [87, p. 219].

These two statements can be compatible only on assumption that the volume of the hydrogen atom is twice as big as that of the oxygen atom. This assumption did not contradict other considerations of Dalton. In spite of this, however, Friedrich Wolff, who translated Dalton's textbook into German in 1812, changed '1 hydrogen' to '2 particles of hydrogen' [89]. Presumably, the cause of such discrepancy was due to the results of the investigation of L. J. Gay-Lussac and F. Humboldt, published in 1809, according to which the volume of hydrogen must be twice as big as that of oxygen when both these elements are wholly combined with one another to form water. The significance of this discovery will be discussed further on in this chapter.

Eventually, in the third chapter of his textbook, entitled 'On Chemical Synthesis', Dalton formulated principles on which he based his considerations concerning the composition of the ultimate particles of various compounds:

Chemical analysis and synthesis go no farther than to the separation of particles one from another, and to their reunion. No new creation or destruction of

3.4. CORPUSCULAR VIEWS OF DALTON

matter is within the reach of chemical agency. (...) In all chemical investigations it has justly been considered an important object to ascertain the relative weights of the simples which constitute a compound. But unfortunately the enquiry has terminated here; whereas from the relative weights in the mass, the relative weights of the ultimate particles or atoms of the bodies might have been inferred, from which their number and weight in various other compounds would appear, in order to assist and to guide future investigations and to correct their results. Now it is one great object of this work, to show the importance and advantage of ascertaining the relative weights of the ultimate particles, both of simple and compound bodies, the number of simple elementary particles which constitute one compound particle, and the number of less compound particles which enter into the formation of one more compound particle.

If there are two bodies, A and B, which are disposed to combine, the following is the order in which the combinations may take place, beginning with the most simple: namely,

1 atom of A + 1 atom of B = 1 atom of C, binary,
1 atom of A + 2 atoms of B = 1 atom of D, ternary,
2 atoms of A + 1 atom of B = 1 atom of E, ternary,
1 atom of A + 3 atoms of B = 1 atom of F, quaternary,
3 atoms of A + 1 atom of B = 1 atom of G, quaternary, etc.

The following general rules may be adopted as guides in all our investigations respecting chemical synthesis.

1st. When only one combination of two bodies can be obtained, it must be presumed to be a binary one, unless some cause appear to the contrary.

2nd. When two combinations are observed, they must be presumed to be a binary and a ternary.

3rd. When three combinations are obtained, we may expect one to be a binary, and the other two ternary... [87, p. 123].

In the early 19th century, many textbooks make a division of substances into simple, double, triple, etc., depending on the number of elements which make up a given substance. As can be seen, in accordance with this tendency, Dalton introduces the names of double, triple, and quadruple atoms. However, it follows from the above-quoted rules that the Daltonian atom of sulphurous acid (SO) would be a double atom and the atom of sulphuric acid (SO_2) a triple atom.

Let us note, however, that Dalton considers only cases where double atoms are formed from two atoms of a different kind. He believes that atoms of the same kind have identical 'sheaths' of the heat atmosphere, and thereby they are not able to have contact with one another. It will become apparent when we take into consideration the view popular in the early 19th century that the element was the limit of chemical analysis, and it was synonymous with the term 'atom'.

Analysing the influence of Dalton's views on the development of chemistry, we can distinguish a few problems which the English scientist advanced or which resulted from his research:

(1) Referring to both single atoms and more complex combinations as atoms and ultimate particles calls for a precise definition of what Dalton understood by the term 'atom' which was after all treated as a synonym of the Newtonian 'ultimate particle'.

(2) What did the Daltonian weights of the 'ultimate particles' signify?

(3) We may wonder whether Dalton's speculations were the final proof of the corpuscular texture of matter.

(4) Dalton's law of multiple proportions, discussed above, concerning the principles of formation of compounds by the elements, contributed to the overthrow of Berthollett's view in the 19th century which claimed that the elements combine with one another in arbitrary proportions.

These problems will be considered in the next sections of this chapter.

3.5. Atomistic Views of Dalton

When analysing Dalton's writings, we come to a conclusion that he never referred to his 'ultimate particles' as indivisible in the sense he is now commonly credited with. Considering his argumentation we may ask how he understood the very term 'atomic theory'. From the very beginning of his scientific activity he employed corpuscular conceptions, but as it follows from his early publications, he realized them fully as late as 1804 while developing the models of the structure of marsh gas (methane), whose molecule, he thought, consisted of one atom of carbon and two atoms of hydrogen; as well as the structure of gaseous ethylene whose molecule consisted, in his view, of one atom of carbon and one atom of hydrogen. In the lectures he delivered in 1810, Dalton stated explicitly that he had arrived at his atomic theory on the basis of physical experiments by considering the structure of gases and liquids and then employed it in chemistry in order to account for the composition of chemical compounds taking into consideration the law of multiple proportions.

In turn, as he writes in a letter to Berzelius, dated 20 September 1812, this theory enabled him to understand the law of multiple proportions, which earlier seemed to him rather puzzling, just as Kepler's laws—in his opinion—appeared puzzling before they were explained by Newton [80, p. 159].

It seems that Dalton did not identify the corpuscular theory with the atomic theory, as is frequently done today, but he associated it rather with the structure of the ultimate particles of chemical compounds. In the same way Auguste Laurent understood the atomic theory nearly half a century later [90, pp. 331–380].

3.6. Views on the Divisibility of the Atom in the 19th Century. Chemical Equivalents

We have noted earlier that the term 'atom' which was used by Dalton cannot be understood unequivocally if we refer only to his publications. However, the text of his lecture delivered at the Royal Institution on 30th January 1810 and his notes related to that lecture enable us to gain a better insight into the meaning which he associated with this term.

> I have chosen the word atom to signify these ultimate particles, in preference to particle, molecule, or any other diminutive term, because I conceive it is much more expressive; it includes in itself the notion of indivisible, which the other terms do not. It may be said that I extend the application of it too far, when I speak of compound atoms; for instance, I call an ultimate particle of carbonic acid a compound atom. Now, though this atom may be divided, yet it ceases to be carbonic acid, being resolved by such division into charcoal and oxygen. Hence I conceive there is no inconsistency in speaking of compound atoms, and that my meaning cannot be misunderstood [80, p. 111].

Thus, Dalton distinguished single, double, triple atoms, etc., but he did not reject the Netwonian conception of ultimate particles. In the notes to the lecture mentioned above we find a rewritten fragment of

Newton's 'Opticks', which has been quoted earlier. Nonetheless, he did not identify his atoms—even the single ones—with the Newtonian particles. Evidence for that is found in a further part of his lecture:

> I should apprehend there are a considerable number of what may be called elementary principles, which never can be metamorphosed, one into another, by any power we can control. We ought, however, to avail ourselves of every means to reduce the number of bodies or principles of this appearance as much as possible; and after all we may not know what elements are absolutely indecomposable, and what are refractory, because we do not apply the proper means for their reduction [80, p. 112].

Dalton (a Quaker preacher) believed, like Newton, essentially that indivisible particles exist; however, he did not state that particles which could not be divided in his time, i.e. the ones which he called single atoms, are in fact ultimately indivisible particles. For Dalton an atom or an ultimate particle of a substance is its smallest amount which still preserves the characteristic properties of that substance. This particle may indeed be decomposed by chemical methods into smaller particles which will, however, reveal different properties. Thus the Daltonian term 'atom' refers both to present-day 'atoms' and to 'molecules'. It should be stressed once again that Dalton was not entirely sure whether atoms, which he regarded as simple atoms, in future would turn out to be complex atoms, although—admittedly—nowadays we still consider as elements all the substances which Dalton considered as elements. It follows from the foregoing discussion that Dalton's conception of atom referred rather to the present-day term 'molecule' than to the 'atom'.

We can see thus that those 'impermeable nuclei', placed by Dalton in the middle of gas particles, are in fact both single and compound atoms or, more generally according to the present terminology, molecules. This is also shown by the illustrations of gaseous substances in Dalton's textbook (Fig. 13). Thus, the 'nuclei' of the hydrogen, nitrogen, and oxygen particles drawn by Dalton as single particles are in fact diatomic molecules, which neither Dalton nor any of the contemporary scientists could decompose. On the basis of Dalton's theory it was not possible to conceive of an ultimate particle—the atom—composed of two identical simple atoms.

3.6. DIVISIBILITY OF THE ATOM

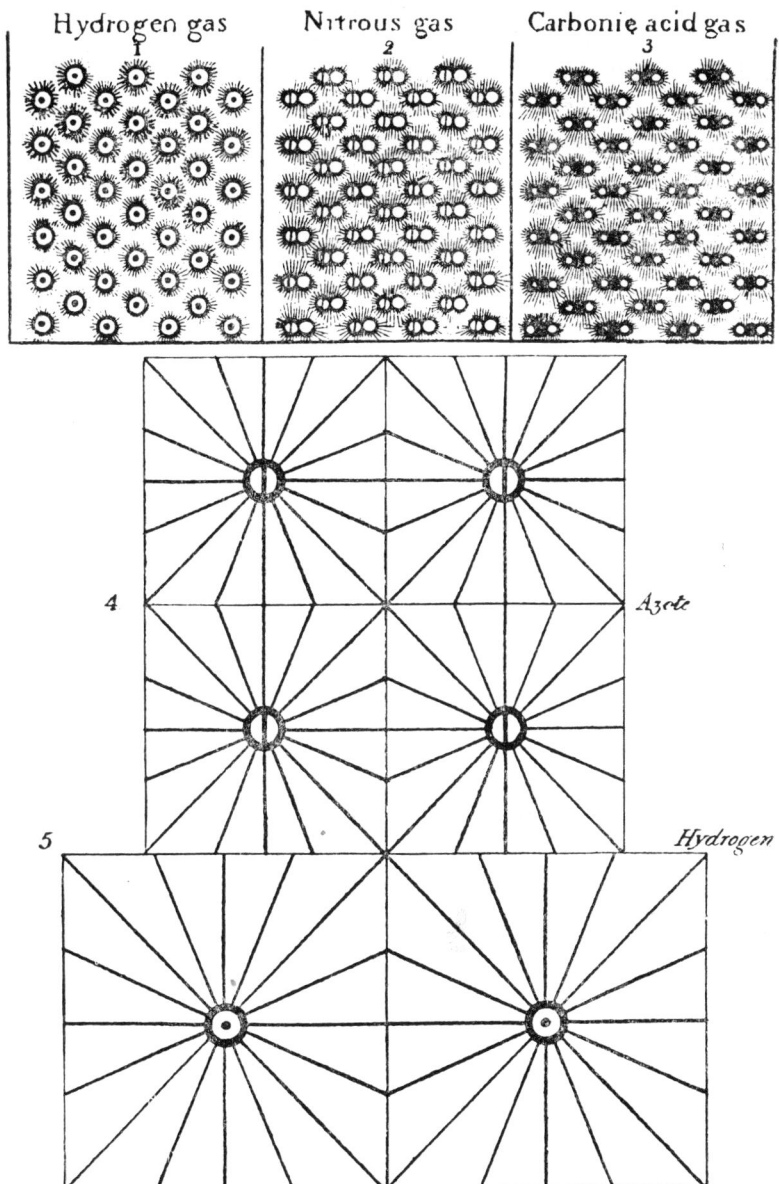

Fig. 13. Plate 7, acc. to [87].

3. THE ELEMENTARY PARTICLE OF MATTER

It is not therefore true that Dalton introduced to science the concept of an 'indivisible' atom. This view may be confirmed by the fact that different outstanding chemists of the 19th century, both those who were active during Dalton's lifetime and the later investigators, predicted the possibility that the atom is a complex form, although they made use of the atomistic terminology and fully appreciated Dalton's contribution. For example, the outstanding Swedish chemist Berzelius, regarded in the first half of the 19th century as the greatest

Fig. 14. Jöns Jacob Berzelius, acc. to[8].

3.6. DIVISIBILITY OF THE ATOM

authority in chemistry, states in a private letter to Dalton in 1812 (published in the *Annals of Philosophy* a year later):

Hence a compound atom is formed by the juxtaposition of several elementary atoms; I suppose likewise that atoms are all spherical, and that they have all the same size [91].

In another fragment of this text we find a more precise definition of what he means by the term divisibility. Berzelius writes:

A compound atom, for very obvious reasons, cannot be considered as spherical; but as it is composed of atoms mechanically indivisible, or which cannot be separated by mechanical means, the compound atom is justly as completely mechanically indivisible as the elementary atom.

Next, Berzelius makes a classification of atoms which might be compared with his classification of bodies discussed in the preceding chapter:

We may divide the atoms into two classes; 1. elementary atoms; 2. compound atoms. The compound atoms are of three different species; namely, 1. Atoms formed of two elementary substances united. We shall call them compound atoms of the first order. 2. Atoms composed of more than two elementary substances; and as these are only found in organic bodies, or bodies obtained by the destruction of such organic matter, we shall call them organic atoms. 3. Atoms formed by the union of two or more compound atoms, as, for example, the salts. We shall call them compound atoms of the second order.

In the first fragment of the above quotation Berzelius stresses the inability to divide an atom merely by mechanical means. We can similarly interpret the definition of the atom provided by Maxwell in the journal *Nature* 60 years later in 1873. He stated: "An atom is a body that cannot be cut in two" [92].

However, the very atomic theory raised doubts among many investigators. The English philosopher Whewell, whose modification of the inductive-deductive reasoning was discussed in Chapter 1, formulated his attitude towards the atomic theory:

The Theory, when dogmatically taught as a physical truth, asserts that all observable quantities of elements are composed of proportional numbers of particles which can no further be subdivided; but all which observation teaches us is, that if there be such particles, they are smaller than the smallest observable quantities. In chemical experiment, at least, there is not the slightest positive evidence for the existence of such atoms. The assumption of indivisible particles, smaller than the smallest observable, which combine, particle with particle, will explain the phenomena; but the assumption of particles bearing this proportion, but not possessing the property of indivisibility, will explain the phenomena at least equally well. The decision of the question, therefore, whether the Atomic Hypothesis be the proper way of conceiving the chemical combinations of substances, must depend, not upon chemical facts, but upon our conception of substance [93].

We have already mentioned that, beginning with an entry in his laboratory diary on 6 September 1803, Dalton denoted the elements by characteristic circles. It was not, however, the first attempt to introduce graphic symbols for the elements. As early as the Middle Ages different symbols were used to denote not only various substances but also chemical operations. Towards the end of the 18th century Adet and Hassenfrantz denoted particular elements by slanted or curved lines, triangles, and circles. Inside the triangles and circles they wrote the initial letters of the Latin names of the elements. It was not until the publication of Dalton's textbook in 1814 that Berzelius proposed a system of a consistent notation which has been used without major modification until the present day:

The chemical signs ought to be letters, for the greater facility of writing, and not to disfigure a printed book. Though this last circumstance may not appear of any great importance, it ought to be avoided whenever it can be done. I shall take, therefore, for the chemical sign, the initial letter of the Latin name of each elementary substance; but as several have the same initial letter, I shall distinguish them in the following manner: 1. In the class which I call metalloids, I shall employ the initial letter only, even when this letter is common to the metalloid and to some metal. 2. In the class of metals, I shall distinguish those that have the same initials with another metal, or a metalloid, by writing the first two letters of the word. 3. If the first two letters be common to two metals, I shall, in that case, add to the initial letter the first consonant which they have not in common: for example, S = sulphur, Si = silicium, St = stibium (antimony), Sn = stannum (tin), C = carboneum, Co = cobaltum (cobalt), Cu = cuprum (copper), O = oxygen, Os = osmium [94].

When two atoms of the same element are in a 'complex atom' of any substance, Berzelius suggests crossing out the symbol of this element with a horizontal stroke at mid-height. The number of oxygen atoms connected with a given element was also denoted by dots, and

3.6. DIVISIBILITY OF THE ATOM

the sulphur atoms by commas placed over the symbol of the element. Berzelius' symbols were not readily adopted by chemists. Even Berzelius himself did not use them in his textbook published in Germany in 1827.

The problem of the sizes of atoms played an important role in the atomic theory. As we have shown, Dalton considered that problem independently of the problem of relative weights of ultimate particles. At some stage of his considerations, he assumed that sizes are equal for all the ultimate particles. This would lead to the conclusion that equal numbers of particles always exist in equal volumes of all gases under identical conditions. And, as we have shown, since the Daltonian ultimate particles are in fact molecules and not atoms, the above statement would be identical with the now generally accepted law of Avogadro, published in the years 1811–1814. Unfortunately, on the basis of other premises, Dalton rejected the concept of equal numbers of particles in equal volumes. Moreover, his theory did not assume the feasibility of the combination of two particles of the same kind unless the compound contained particles of other elements.

The above standpoint of Dalton, generally known and accepted by a considerable fraction of the contemporary chemists, hampered an understanding of the next quantitative law—i.e. the law of simple combining volumes—which Gay-Lussac presented at a public meeting on 31 December 1808 and published in the following year. He developed this law on the basis of his own investigations and the experiments of different chemists; amongst others, in 1805 together with Alexander von Humboldt he noticed that 100 volume parts of oxygen after combining with 200 volume parts of hydrogen gives 200 volume parts of steam.

In the above-mentioned memoir Gay-Lussac stated:

(...) it is my intention to make known a new property in gases, the effects of which are regular, by showing that these substances combine amongst themselves in very simple proportions, and that the contraction of volume which they experience on combination also follows a regular law (...). Suspecting, from the exact ratio of 100 of oxygen to 200 of hydrogen which M. Humboldt and I had determined for the proportions of water, that other gases might also combine in simple ratios, I have made the following experiments. I prepared fluoboric, muriatic, and carbonic gases, and made them combine successively with ammonia gas. 100 parts of muriatic gas saturate precisely 100 parts of ammonia gas, and the salt which is formed from them is perfectly neutral, whether one or other of the gases is in excess [95].

Next Gay-Lussac stated that 100 parts of nitrogen can combine with 50, 100, or 200 parts of oxygen to yield different gases (different nitrogen oxides), and that "ammonia gas consists of three volumes of hydrogen and one volume of nitrogen". (It should be recalled that according to the principles proclaimed by Dalton a complex atom of ammonia was believed to consist of one hydrogen atom and one nitrogen atom). Gay-Lussac writes further:

(...) We can see from different examples that contraction, to which two gases are subjected while mixed, remains in a close relationship with their volume.

In spite of a contradiction between the above law of simple combining volumes of reacting gases and the conclusions resulting from the atomic theory of Dalton, the views of both investigators were simultaneously adopted and recognized by the majority of the outstanding chemists of the 19th century. For example, Berzelius, as can be seen from a subsequent part of the quoted publication of 1813, valued the law of combining volumes even higher than the atomic theory. He wrote:

Hence there is no other difference between the theory of atoms and that of volumes, than that the one represents bodies in a solid form, the other in gaseous form. It is clear, that what in the one theory is called an atom, is in the other theory a volume. In the present state of our knowledge the theory of volumes has the advantage of being founded upon a well constituted fact, while the other has only a supposition for its foundation. In the theory of volumes we can figure to ourselves a demi-volume, while in the theory of atoms a demi-atom is an absurdity [94].

As can be seen from the foregoing, the volume of the combining elements, which were then frequently identified with the ratios of the numbers of the combining atoms, did not conform with the weight ratios of those elements. At that moment, W. H. Wollaston published his study which we have mentioned earlier. The author proposed a scale:

(...) designed to (...) directly indicate the actual weights of the several ingredients, contained in any assumed weight of the salt under consideration,

3.6. DIVISIBILITY OF THE ATOM

and also the actual quantities of several reagents that may be used, and of the precipitates that would be attained by each. In the information of this scale, it is requisite in the first place to determine the proportions in which the different known chemical bodies unite with each other and express these proportions in such terms that the same substance shall always be represented by the same number.

It is to Richter* that we are originally indebted for this mode of expression, for having first observed that law of permanent proportions on which the possibility of this numerical representation is founded [96, p. 2].

And further:

According to Mr. Dalton's theory (...) chemical union in the state of neutralization takes place between single atoms of the substances combined; and in the case where there is a rebundance of either ingredient, then two or more atoms of this kind are united to only one of the other. According to this view, when we estimate the relative weights of equivalents, Mr. Dalton conceives that we are estimating the aggregate weights of a given number of atoms and consequently the proportions which the ultimate single atoms bear to each other. But since it is impossible in several instances, where only two combinations of the same ingredients are known, to discover which of the two compounds is to be regarded as consisting of a pair of single atoms, and since the decision of this question is purely theoretical, and by no means necessary to the formation of a table adapted to most practical purposes, I have not been desirous of warping my numbers according to the atomic theory, but have endeavoured to make practical convenience my sole guide and have considered the doctrine of simple multiples, on which that of atoms is founded, merely as a value assistant in determining, by simple division, the amounts of those quantities that are liable to such definite deviations from the original law of Richter [96, p. 7].

Like Thomas Thomson, Wollaston denoted equivalents in relation to the equivalent of oxygen, for which he assumed the value 10. Later chemists, e.g. Berzelius, assumed as a rule the value 100 for oxygen. Since oxygen was taken as a reference element, the numerical values of equivalent weights were twice as big as those of atomic weights, contrary to the present practice, where the atomic mass is divided by valency. It follows from the above discussion that the number which Dalton had given as 'atomic weights' were rather equivalent weights,

---

* Wollaston ascribes the law of constant composition to Richter and not to Proust (R.M.).

and this refers both to the elements (simple atoms) and compounds (complex atoms). From Wollaston's time the quantitative ratios of the elements in chemical compounds were defined as either equivalent ratios or volume ratios of quantities which were believed to be proportional to the number of atoms. Similarly, chemical formulae used in the 19th century reflected either equivalent ratios or volume ratios, i.e. the number of atoms. For example, J. Seweryn Zdzitowiecki, in his textbook of chemistry published in Warsaw in 1850, provides two formulae for water: an equivalent formula—HO, and an atomic formula—H_2O [97]. Further, this author, like the majority of the authors of inorganic chemistry textbooks until the 1800s, rather employs equivalent formulae, whereas atomic formulae occurred more often in the textbooks of organic chemistry in the same period.

In the year 1815, William Prout, "basing mainly on the doctrine of volume generalized for the first time by M. Gay-Lussac, and which is generally adopted by the chemists" [98], calculated the 'atomic weights' of 59 elements, oxides, hydrides, and also metals with respect to the atomic weight of hydrogen recognized as a unit beside the weights calculated with respect to the atomic weight of oxygen (10). With the exception of four cases, the weight values calculated by Prout with respect to hydrogen were whole numbers. For example, we find the following values of these weights in his tables: oxygen—16 or 8, nitrogen—14, carbon—6, sulphur—16, chlorine—36, water—9, carbon oxide—14, atmospheric air (regarded as a compound)—14.4, chromic acid—76, hydrochloric acid (HCl)—18.5, ammonia—8.5, aluminium—8, mercury—100, gold—200. The above-quoted fractional values, resulting from Prout's own calculations, as well as from the measurements of other investigators were regarded by Prout as an outcome of errors in the experiments; he was convinced that all correctly performed measurements must yield integral number values of 'atomic weights'.

In the next year, the English investigator considered the problem of the ratio of the specific weight of the elements to their atomic weight and generalized his observations in the following way:

If the views we have ventured to advance be correct, we may almost consider $\pi\varrho\omega\tau\eta$ $\upsilon\lambda\eta$ (*prote yle*) of the ancients to be realised in hydrogen; an opinion, by the by, not altogether new. If we actually consider this to be the case, and further

3.6. DIVISIBILITY OF THE ATOM

consider the specific gravities of bodies in their gaseous state to represent the number of volumes condensed into one; or, in other words, the number of the absolute weight of a single volume of the first matter (*prote yle*) which they contain, which is extremely probable, multiples in weight must always indicate multiples in volume, and vice versa; and the specific gravities, or absolute weights of all bodies in a gaseous state, must be multiples of the specific gravity or absolute weight of the first matter (*prote yle*), because all bodies in a gaseous state which unite with one another unite with reference to their volume [99].

As can be seen, the famous hypothesis of Prout that all atoms, i.e. including compounds, consist of atoms of hydrogen as units of the ancient *prote yle* was based first of all on the law of combining volumes formulated by Gay-Lussac. The English authors accused Prout that his hypothesis overthrew Dalton's atomic theory. However, Prout, in a letter published in a collection 'On the Atomic Theory' defended his position:

There is no one who can possibly have greater respect for Mr. Dalton, and all he has done, than myself, and I am a firm believer in his principles as far as they go, because I believe them to be founded in truth [100].

We remember, however that Dalton did not state explicitly that the particles which he called atoms must be indivisible; therefore, there is no reason to raise such objections against Prout as the contemporary scientists did.

In 1819, two different methods of determining atomic weights (atomic masses) were proposed. E. Mitscherlich based his method on the isomorphism of some crystals. He stated that two isomorphic crystals, in which only one element was replaced by another, must contain these elements in equal proportions with respect to the other common elements. Thus, on the basis of the isomorphism of Na_2HPO_4 and Na_2HAsO_4 crystals, he was able to determine the atomic mass of arsenic because he knew the relevant value for phosphorus. Another method, employed frequently in the subsequent years, was proposed by P. Dulong and A. Petit. They showed that the so-called 'atomic heat', i.e. the product of the specific heat and atomic mass of a given element, has a constant value independent of the kind of the element. Thus, dividing this value by the measured specific heat one obtains the atomic mass of the investigated element.

3. THE ELEMENTARY PARTICLE OF MATTER

Values of atomic weights obtained by these methods were, however, frequently fractional. Thus, Prout's statement about the whole value of those weights did not withstand the test of time. In the year 1832 the English chemist Edward Turner showed that his analysis of lead chloride and an analysis of silver chloride performed by Berzelius proved that exactly the same value of the atomic weight for chlorine (35.45) was obtained, hence it could not have been a whole value [101]. In the year 1845, a letter of Berzelius, then the greatest authority in chemistry, was published. The Swedish scientist accuses Prout of an arbitrary choice of results on which he based his hypothesis and states that chemists would be able to perform a transmutation of one element into another if this hypothesis were true [102]. Berzelius was not familiar with such facts; however, many precise measurements of atomic and molar masses, performed mainly in his laboratory, gave fractional results. Thus, Berzelius and Turner demonstrated rather convincingly that the premises on which Prout based his hypothesis were fallacious, and consequently it was rejected thirty years after its formulation.

However, neither Dalton's considerations nor the rejection of Prout's hypothesis stopped the research on the divisibility of matter. Numerous outstanding chemists, such as Jean Baptiste Dumas in 1851, John H. Gladstone in 1853, T. Sterry Hunt in 1867, and Joseph Norman Lockyer in 1879, analysed the arithmetic relationships between the atomic and equivalent weights of different elements, and on this basis they attempted to find one or two simple atoms, the combination of which might form all other atoms. At the same time they tried to assume a part of the hydrogen atom, e.g. one half or one fourth, to be the simplest particle. In 1885, Marcelin Berthelot assessed these attempts critically:

(...) it is clear that all numerical conjectures could be acceptable if a common unit of atomic weights were adopted below the limit of errors which we cannot avoid [103].

This remark is of course correct. However, it is good evidence that chemists throughout the 19th century could not give up the idea of the divisibility of the 'indivisibles' i.e. atoms.

3.6. DIVISIBILITY OF THE ATOM

Many works published at that time lead to a similar conclusion. Their authors made a distinction between a chemical atom and the smallest portion of the texture of matter—a physical atom. In the 'Familiar Letters on Chemistry' published for the first time in 1844, the author Justus Liebig, an eminent German organic chemist, writes:

> It is impossible for the human mind to imagine particles of matter to be absolutely indivisible, since they cannot be infinitely small in a mathematical sense, that is to say, altogether without extension because they possess a certain weight; and how minute soever we may assume this weight to be, yet we cannot consider the division of a particle possessing weight to be impossible into two, three, nay into a hundred parts. But we may also suppose that the ultimate atoms of bodies are only physically indivisible; they are only incapable of further subdivision so far as our powers of perception enable us to judge.
>
> A physical atom in this sense, then, is a conglomeration of innumerable smaller imaginary particles, held together by a force or forces more powerful than all the means of our command for their further subdivision or dissolution.
>
> The term atoms is employed by chemists in a sense precisely analogous to that attached by them to the word element (...). Without disputing the infinite divisibility of matter, the chemist merely maintains the firm and immovable foundation of his science, when he admits the existence of physical ATOMS as a truth, entirely incontrovertible [104].

A similar view was expressed in 1867 by August Kekulé:

> I have no hesitation in saying that, from a philosophical point of view, I do not believe in the actual existence of atoms, taking the word in its literal signification of indivisible particles of matter. I rather expect that we shall some day find, for what we now call atoms, a mathematico-mechanical explanation, which will render an account of atomic weight, of atomicity, and of numerous other properties of the so-called atoms. As a chemist, however, I regard the assumption of atoms, not only as advisable, but as absolutely necessary in chemistry. I will even go further, and declare my belief that chemical atoms exist, provided the term be understood to denote those particles of matter which undergo no further division in chemical metamorphoses. Should the progress of science lead to a theory of the constitution of chemical atoms—important as such a knowledge might be for the general philosophy of matter—it would make but little alteration in chemistry itself. The chemical atom will always remain the chemical unit [105].

It will be worth quoting Mendeleev's statement of 1871 contained in the supplement to *Annalen der Chemie und Pharmacie* two years after he published his Periodic System:

3. THE ELEMENTARY PARTICLE OF MATTER

Fig. 15. Dmitri Mendeleev, acc. to [8].

Everyone knows the fate of Prout's hypothesis. It seems to me there is no basis whatsoever to accept this hypothesis. Even if we assume that the matter of the elements is completely homogeneous, there is no reason to assert *a priori* that n weight parts of one element or n atoms, being transformed into one atom of another element, will yield the same n weight parts (...). The law of conservation of weight can be considered as a special case of the law of conservation of force or motion.

Weight can be caused by a special kind of motions of matter, and there is no reason to exclude that possibility of transformation of these motions into chemical energy or any other form of motion during the formation of an elemental atom (...). Expressing this thought I want to say merely that there exists a possibility of reconciling silently by the chemists a long-time cherished opinion about the complex nature of the elements while not accepting Prout's hypothesis [106].

Thus, even Mendeleev, who in our opinion placed atoms in an array of pigeon-holes, did not reject the conception that the formations so arranged are not the smallest particles of matter. He attempted to justify the fractional atomic masses by the law of conservation of mass (i.e. weight—in the terminology used a hundred years ago), which was extended by the law of conservation of energy (i.e. force—in Mendeleev's terminology) and momenta (motions). The fractional values of atomic masses were accounted for by Mendeleev as a possibility of the emission of the energy as a result of the combination of the constituent parts of an atom into a new form. Although this idea reminds us of Einstein's conception about the equivalence of mass and energy, it was based on a false premise because fractional masses, known in the time of the formulation of the Periodic System, were a statistical result of the isotopic composition of the elements (of which Mendeleev was unaware, of course), and not the result of the defect of mass, as the above-quoted fragment might suggest.

The turn of the 19th century brought an experimental confirmation of the thesis of the divisibility of atoms. We shall return to these problems further in this chapter.

3.7. The Fate of Avogadro's Hypothesis

As we have mentioned, many chemists in the early 19th century did not see any contradiction between Dalton's atomic theory and the law of simple combining volumes discovered by Gay-Lussac. However, there were a few who noticed this contradiction, such as Amadeo Avogadro and André Maria Ampère. As early as 1811 Avogadro writes:

M. Gay-Lussac has shown (...) that gases always unite in a very simple proportion by volume, and that when the result of the union is a gas its volume also is very simply related to those of its components. But the quantitative proportions of substances in compounds seem only to depend on the relative number of molecules which combine and on the number of composite molecules which result. It must then be admitted that very simple relations also exist between the volumes of gaseous substances and the numbers of simple or compound molecules which form them. The first hypothesis to present itself in this connection, and apparently even the only admissible one, is the supposition that the number of integral molecules in any gases is always the same for equal volumes, or always proportional to the volumes (...). (...) the volume of water in the gaseous state is, as M. Gay-Lussac has shown, twice as great as the volume of oxygen which enters into it. (...). But a means of explaining facts of this type in conformity with our hypothesis presents itself naturally enough: we suppose namely, that the constituent molecules of any simple gas (...) are not formed of a solitary elementary molecule but are made up of a certain number of these molecules united by attraction [107].

Thus Avogadro distinguished total molecules (molecules of a compound), simple or constituent molecules (single atoms of an element which forms a compound), and elemental molecules (molecules of a pure element). Similarly, in 1814 Ampère emphasized that the number of molecules must be proportional to the volume of gas.

Nonetheless, in spite of Ampère's support, Avogadro's hypothesis could not be recognized as true in the early 19th century. It postulated that molecules consisting of atoms of the same kind exist in a gas, and according to the contemporary view this was not plausible. This view was based on Daltons' atomic hypothesis which has been discussed above and also on the electrochemical theory of the structure of chemical compounds developed by Berzelius. It will be discussed in detail in the next chapter. Thus by the mid 19th century chemists still did not distinguish the modern concepts of atom and molecule, and as late as 1858 Justus Liebig wrote about an atom of grape sugar made of an atom of carbonic acid in which one oxygen atom was substituted by one hydrogen atom. It was still believed that compounds were composed primarily of one atom, or rather an equivalent of each kind.

After the rejection of Avogadro's hypothesis in the research on mineral compounds a molecule of hydrogen H_2 was assumed to be a unit of atomic weight because it was regarded as one atom. On the other hand, in organic chemistry, which was then rapidly developing, the exchange reaction of hydrogen in hydrocarbons with other atoms or groups of atoms was the basis of determining the atomic composition and molar masses of compounds. Molar masses were thus referred to the mass of an individual hydrogen atom. As a result the chemical formula of the same compound could have a different form depending on whether it was determined by the methods of inorganic or organic chemistry. This created an irreducible gap between both branches of chemistry; it was sometimes felt that 'mineral' chemistry

3.7. THE FATE OF AVOGADRO'S HYPOTHESIS

was a quite different science from organic chemistry. This divergence was even greater due to the fact that organic chemists employed different theories of the structure of compounds. These will be discussed in the next chapter. In consequence, 16 different formulae of acetic acid were given simultaneously by different chemists in 1860. The atomic hypothesis seemed quite unnecessary to organic chemists; in 1834 the French organic chemist Dumas exclaimed: "If I were God I would obliterate the word atom from science".

A way out of this confused state of chemistry at the beginning of the second half of the last century was shown by Avogadro's countryman, Stanislao Cannizzaro. In 1858, he published in *Nuovo Cimento* an open letter to Professor S. de Luca entitled *Sunto di un Corso di filosofia chimica* in which it was pointed out that the discrepancies (discussed above) could be avoided if the measurements of atomic weights were based on Avogadro's hypothesis. It held that molar masses should be proportional to the density of a substance in the gaseous state. Based on this conclusion, and assuming that the molar mass of hydrogen equals 2 (since a molecule of gaseous hydrogen is composed of two atoms), Cannizzaro compiled a table of the molar masses of 33 elements and compounds. On this basis he came to the conclusion that:

(...) different amounts of one and the same element making up different molecules are the complete multiples of one and the same amount which always occurs indivisibly and which, with some justification, should be called an atom.

Thus, on the basis of the investigation of gases, Cannizzaro considered an atom to be "such an amount (of the element—R.M.) which is always included as a whole in equal volumes of both a free body and its compounds" [108, p. 12].

The above quotations show that Cannizzaro thought it necessary to make a clear distinction between the concepts of atom and molecule. Earlier, the two French chemists August Laurent and Charles Gerhardt had reached a similar conclusion drawn from their consideration of the structure of organic compounds. However, they did not dare to present their conception quite as explicitly so long as Berzelius was alive, because he was a dedicated opponent of Avogadro's hypothesis.

Cannizzaro's publication did not initially meet with a wide response. Neither did his contribution to the First International Congress of Chemists in Karlsruhe in 1860 convince the participants, although the objective of the Congress was to bring order into the confusion in chemical concepts. Cannizzaro, however, distributed reprints of his open letter to Professor Luca among the participants. After careful reading of this publication, the relevance of his theses was generally accepted, particularly with reference to molecules in a gaseous state. It was agreed that two identical atoms can combine with each other to form one molecule, and that the terms atom and particle (German *Teilchen*) are not to be regarded as synonymous. However, many investigators from that period considered atoms and molecules only as terms which served as a convenient representation of phenomena and not as real forms: only volumes of bodies were real for them. In a lecture delivered in London in 1872 in honour of Faraday, Cannizzaro defended the reality of atoms and molecules, basing himself not only on Avogadro's and Ampère's publications but also on the kinetic theory of gases formulated anew in the years 1856–1858 by August Krönig and Rudolf Clausius. Cannizzaro proved that the use of unitary volumes was only a convenience, because weight (i.e. mass) is the only quality which is inherently associated with this concept and characterizes the substance.

Based in great measure on his own denotations of the atomic masses of the elements and the molar masses of compounds (from the measurements of the density of gases), Cannizzaro stressed the significance of atomic and molar masses, in spite of a fairly common opinion at that time which attributed the essential importance to chemical equivalents. Many investigators of that period regarded molar masses and equivalents as identical values. Recognition of a difference between these values was a necessary condition for the future development of chemistry because, for example, the equivalent of calcium is 20, and it atomic mass is 40. Thus the conviction of the essential significance of atomic masses permitted the completion of the work on the systematization of the elements. Another important factor which favoured this work was the clarification of the concept of valency around the year 1860. The development of this concept will be discussed in the next chapter.

3.8. Classification of the Elements

A tendency to classify the known substances had already existed in antiquity. The classification was based on certain external physical features and occasionally on the site where a substance was found. In the Middle Ages the alchemists classified substances on the basis of their chemical properties. Caustic substances were obtained from non-caustic substances, generally called salts. In the process of calcination—as we have already said in Chapter 2—volatile 'spirits' were given off by these substances and, after dissolution in water, they became acids. The calcination residue was a caustic solid body, called *alkali* since Arab times. Alkalis were also obtained by leaching plant ashes with water and vaporization of water from the solution obtained. A solution of acetic acid (vinegar) was obtained through fermentation. Various sour plant juices were known besides. Acid was distinguished by taste and alkalis were identified by the fact that acids poured on them caused effervescence, i.e. release of 'air'. It was thus realized that salts consist of acidic and alkaline parts. In 1668, O. Tachenius wrote: "... everything that is salty can be divided into two substances: some alkalis and acid which dissociated and then again associated yield salt" [45, p. 11]; and further he wrote: "all salts which neutralize acids are referred to by one general name Alkalis which can be found not only in the family of plants but also in that of minerals and animals".

From the 17th to the 19th centuries substances were divided according to the families or kingdoms of their origin into a mineral, plant, or animal family. In consequence, the same chemical compound—as we know today—but obtained from different sources, was sometimes regarded as a different substance. Substances derived from minerals were believed to be the simplest, whereas those derived from plants appeared to be complex, and animal-derived substances were regarded as the most complicated.

3. THE ELEMENTARY PARTICLE OF MATTER

In 1665, in a treatise entitled *Experimenta et considerationis de coloribus* Robert Boyle provided a new way of distinguishing acids and alkalis on the basis of the colour of the added solution of some vegetable dyes, such as *Syropum Violarum*, extracts of violets, cornflowers, cochineal, and the solution of litmus. Boyle's publication started the application of indicators in chemistry.

Another criterion of classification was appearance and physical properties. For example, all crystalline substances that are soluble in water were included in one group—salt—regardless of their chemical properties. Thus, sugar and tartaric acid were regarded as salts—and as we have seen from Tachenius's quotation—salt was a very general term which included many non-metallic solid substances. This criterion was employed more extensively by the French chemist Etienne François Geoffroy who tried to systematize substances according to their increasing affinity with respect to a few selected chemical substances. Geoffrey's table will be discussed in Chapter 5. Another criterion was employed by Lavoisier in his table of elements, which we have presented in the previous chapter. The elements were arranged in particular groups according to the properties of their compounds or, more precisely, the acidic and basic properties of their oxides. The third chemical criterion can be found in Richter's tables, containing substances arranged according to their chemical equivalents.

In 1809, Thomas Young attempted to improve Geoffrey's criterion. He determined, with an accuracy up to one thousandth, the 'forces of attraction' of different metals exerted by particular acids, and on this basis he systematized them. For example, assuming the 'force of attraction' of baryte earth (regarded as an element) by sulphuric acid to be 1000; he attributed the value 894 to the force of attraction of potash by this acid and 801 to potash attracted by phosphoric acid [109].

A similar approach was taken by Ampère, who proposed an interesting classification of the elements:

> I shall confine myself in this paper to the simple bodies, and shall divide them into three heads. I shall offer in the first, some general considerations on the order according to which it is proper to arrange bodies, so that this order may be as conformable as possible to their natural analogies; and on the means of

3.8. CLASSIFICATION OF THE ELEMENTS

avoiding the junction which has been hitherto made of the metals with bodies very different in almost all their other characters, and which have only been brought together because the energy of their affinity for oxygen is nearly the same (...).

Under the second head, I shall unite under natural genera the bodies which present characters of resemblance so multiplied and important that it is impossible to separate them in every classification which shall not be purely artificial (groups of such metals as iron, nickel, cobalt—R.M.) (...)

The last head of this paper will have for its object to examine once more the various genera into which all these bodies shall have been distributed (...) in order to assign to each of them a distinctive character formed by the union of some remarkable properties, chosen in such a way that they cannot be found at once but in a body appertaining to the genus which it is wanted to characterize, and to see at the same time according to what principles of nomenclature we could, if necessary, establish for each genus a denomination common to all the bodies which form part of it [110].

Further in his work, Ampère divided the elements into three classes:

1. Gasolites, i.e. substances which can form stable gases when combined with one another in the same class.

2. Leukolites, i.e. metals which melt at a temperature below 25° Wedgewood, their oxides forming colourless solutions with colourless acids.

3. Chroikolites, i.e. metals which melt at temperature above 25° Wedgewood, their oxides forming coloured solutions with colourless acids.

Ampère subdivided each of these classes into several types so that each type included two to five elements which shared one feature. For example, only carbon and hydrogen belonged to the type anthracites in the class of gasolites as "bodies which combine with one element of air when exposed to it at a sufficient temperature". The type agrirides in the class of leukolites included bismuth, mercury, silver, and lead; while tephralides included sodium and potash. Ampère's classification was unclear, the particular types included so few elements that it did not actually constitute a practical system. Now this classification is only a historical curiosity.

In the same year of 1817, Wolfgang Döbereiner attempted to find an arithmetic relationship between the atomic weights or equivalents of bodies which have similar properties. He discovered that the equivalent weight of strontium oxide (50) is an arithmetic mean of the equivalent weights of calcium oxide (27.5) and barium oxide (72.5). In 1829, Döbereiner observed further triada of elements with similar properties. For example, he found that the atomic weight of the middle atom is an arithmetic mean of the atomic weights of the outside atoms in the following triads: Cl Br I, Li Na K, P As Pb, S Se Te. He also found such a 'triadic' dependence in some atoms which do not have similar properties but whose equivalent weights increase in the following sequence: C N O, Fe Cr Mn, Ni Cu Zn, Pt Ir Os. A similar conception of triads was developed by Jean Baptiste Dumas in 1851. He was often quoted by the later 'classifiers'. The basis of formation of the groups of elements for him was a comparison of the solubility of their compounds in acids [111].

In the year 1843, Leopold Gmelin devised a systematic arrangement of the elements in which he attempted to take account of both their arrangement according to their increasing equivalent weights and the concept of chemical groups. In view of the great uncertainty in the values of equivalent masses, this system is not clear enough for us; however, it enabled John Gladstone, ten years later, to state that an increase in the atomic weight of the elements belonging to the particular groups can be expressed by a formula $a+nx$, where n assumes the values 0, 1, 2, 3, respectively. A similar suggestion was proposed by Max Pettenkoffer as early as 1850. He proposed adding a multiple of eight to the definite weight of a lighter element, hence x would equal 8.

In the year 1854, the American chemist Josiah Parson Cooke employed an arithmetic method in order to divide the elements into six series: the weights of atoms differed by 3, 4, 5, 6, 8, or 9, depending on their placement in the relevant series. He showed that the elements that were in one series had certain similar properties, e.g. their compounds had an analogous crystallographic form. However, a few elements, e.g. oxygen and manganese, belonged to three series simultaneously [112].

Such was the state of art as far as the classification of the elements is concerned in 1860 when the International Congress of Chemists was held in Karlsruhe. After this Congress, as we have said, the

3.8. CLASSIFICATION OF THE ELEMENTS

concepts of the atom and molecule were defined more precisely, and on this account chemists could develop more uniform views on the question of the atomic weights of particular elements. In the years 1862–1864 four investigators proposed new conceptions of the classification of the elements.

In the years 1862 and 1863, the French geologist Alexandre Emile de Chancourtois devised a technique for the classification of the elements which he presented in a work entitled *Vis tellurique*, published in 1863. We shall quote a fragment of this publication since it reflects the approach taken by the chemists over a hundred years ago, and the prevailing confusion:

Geological studies in the field of research opened up by M. Elie de Baumont (...) have led me, for completion of a lithological memoir on which I am now engaged, to a natural classification of the simple bodies and radicals by a table in the form of a helix, founded on the use of numbers which I call characteristic numbers or numerical characteristics.

My numbers, which are immediately deduced from the measure of the equivalents or other physical or chemical capacities of different bodies, these being reduced to half in the case of hydrogen, nitrogen, fluorine, chlorine, bromine, iodine, phosphorus, arsenic, lithium, potassium, sodium and silver; in other words I either divide the equivalents of these bodies by two in the system in which oxygen is taken as 100, or multiply by two equivalents of other bodies in the system in which hydrogen is taken as unity [113].

Thus Chancourtois converts the known equivalent weights to atomic masses. Chancourtois drew a helical line on a cylinder in such a way that they crossed its generatrices at an angle of 45° and then he marked on it the characteristic numbers of the elements assuming as a unit of length on the perimeter one full turn, which corresponded to the characteristic number of oxygen, or 1/16 of a turn, which corresponded to the characteristic number of hydrogen. After developing the surface of the cylinder, Chancourtois stated:

The relations between the properties of different bodies are manifested by simple geometrical relations between the positions of their characteristic points. For instance, oxygen, sulphur, selenium, tellurium fall approximately on the same generating line, while calcium, iron, strontium, uranium and barium fall on the opposite generating line (...). Each helix drawn through two characteristic points and passing through several other points or near them, brings out relations of a certain kind between their properties.

Chancourtois thought that his screw might explain, among other things, the spectrum colours of the elements. Furthermore, it reflected the fact that the differences between the characteristic numbers of the elements which were on the same generatrix are multiples of 16. Thus Chancourtois was the first 'classifier' who succeeded in relating the arrangement of the elements according to their increasing atomic masses to the existence of the families of elements with similar properties. However, since his publication did not contain any diagrams, it was not noticed for a long time.

On the other hand, two articles by John A. R. Newlands under a common title: 'On Relations Between the Equivalents' published in the *Chemical News* on 7 February 1863 and 30 July 1864 met with a wider response. In the first article Newlands classifies the elements only according to their chemical properties, in contrast to his predecessors who, in great measure, also took account of physical properties. On this basis he distinguishes eleven groups of elements, six of which correspond to the main groups of the subsequent periodic system, i.e. alkalic elements, elements of alkalic earths, halogens, chalcogens, the nitrogen and the carbon families. In the second article Newlands wrote that in the fascicle of the 2nd day of this month one of the correspondents subscribed as 'Studiosus' pointed to the existence of the law concerning the phenomenon that the atomic weights of elementary bodies are, with a few exceptions, either the multiples of eight or nearly the multiples of eight [114].

Next Newlands shows that, according to the division into groups which he presented in a previous publication, it is possible to speak about the multiples of eight with respect to the differences between the equivalents of the elements which belonged to a given group. In many publications Newlands is regarded as the creator of the law of octaves.

In the same year, 1864, William Odling arranged about sixty of the then known elements in order of increasing atomic weights and found that, despite a few blank spaces, the weights increased fairly regularly [115]. Today we may presume that those blank spaces were caused in part by a wrong determination of atomic weights, and in part by the fact that some elements were not known at that time. Next

3.8. CLASSIFICATION OF THE ELEMENTS 141

Odling arranged his series in five columns so that the elements having similar properties, i.e. ones that Döbereiner arranged in triads, were grouped in one calumn. These columns corresponded to the elements which were placed on Chancourtois' screw. Odling's system had a lot of gaps and inconsistencies and its author failed to draw any relevant conclusions.

Also in 1864 Lothar Meyer, in his textbook *Die modernen Theorien der Chemie*, listed 27 elements according to their increasing atomic weight and arranged them in vertical columns so that elements of similar chemical properties appeared in each column, e.g. elements of equal valencies. He then asserted that the atomic weights of the two lightest elements of each group differ by 16, and those of two further elements by 44, and those of the two heaviest elements by 88. In 1870, he published a full system of the elements which, however, followed the publication of Mendeleev's periodic system.

In Mendeleev's System of the elements we find all the ways of classifying the elements which had been attempted earlier. However, a full account of all those attempts enabled him to formulate quite new conclusions. Mendeleev completed the description of his System on 1st March (17th February according to the Julian calendar) 1869, and in the same year he published it in *Zhurnal Russkovo Khimitcheskovo Obshchestva* and, in the German version, in *Zeitschrift für Chemie*. On account of the importance of that publication and its conciseness, we shall quote *in extenso* from the German translation:

1. The elements, if arranged according to their atomic weights, show an evident periodicity of properties.

2. Elements which are similar as regards their chemical properties have atomic weights which are either of nearly the same value (platinum, iridium, osmium), or which increase regularly (potassium, rubidium, caesium).

3. The arrangement of the elements, or of groups of elements, in the order of their atomic weights, corresponds with their so-called valencies.

4. The elements which are most widely distributed in nature have low atomic weights, and sharply defined properties. They are therefore typical elements, and therefore, justly, the lightest element H has been assumed as a typical scale.

5. The magnitude of the atomic weight determines the character of an element.

6. The discovery of many yet unknown elements may be expected.

7. The atomic weight of an element may sometimes be corrected by the aid

142   3. THE ELEMENTARY PARTICLE OF MATTER

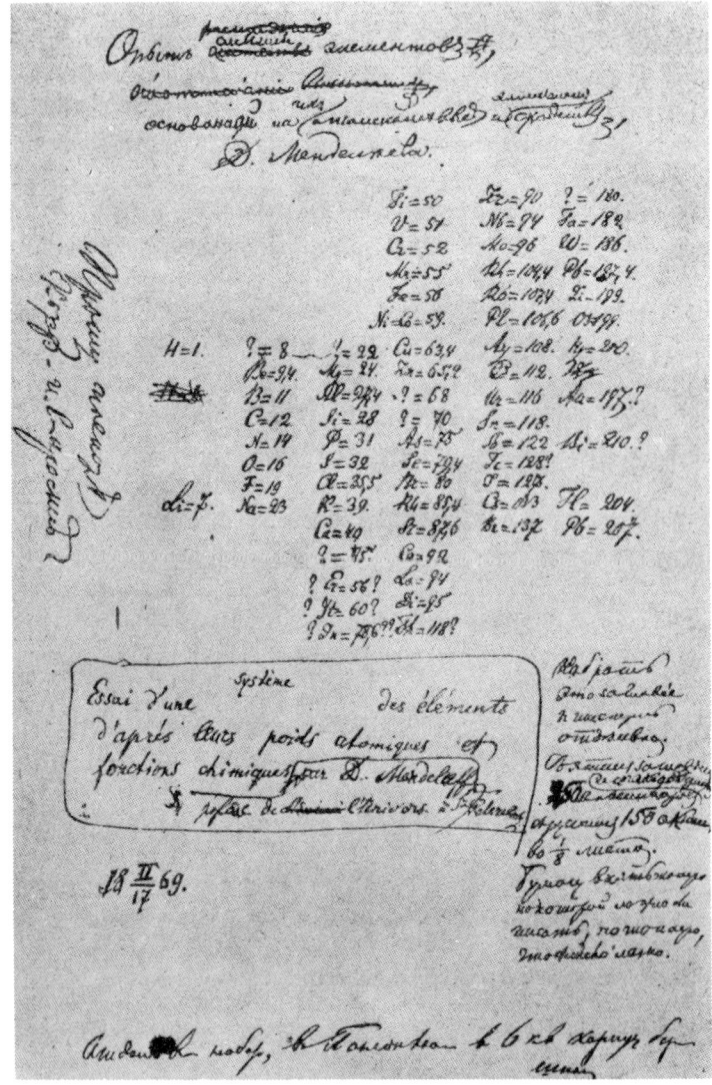

Fig. 16. Manuscript of Mendeleev's Periodic System of the Elements from 1869, from a guide-book to the Dmitri Mendeleev Museum in Leningrad, USSR.

3.8. CLASSIFICATION OF THE ELEMENTS 143

of a knowledge of these adjacent elements, e.g. Te cannot have an atomic weight of 128 but 123–126. New analogies between the elements result from the above table. Thus Bo (borium—R.M.) turns out to be an analog of Al, which, as we know, was established experimentally a long time ago [116].

This statement was supplemented by a table which represented an exact copy of the table made by Mendeleev in his laboratory diary; it is reproduced in Fig. 16.

The last three points of the publication quoted require an additional commentary. A remark contained in point 7 results from the fact that the atomic mass of iodine which, according to its chemical properties, must follow tellurium, is smaller than tellurium by 1. As we know today, the values provided by Mendeleev were in principle correct. The cause of this and a few other 'irregularities' discovered later was understood only as late as the 1930s.

Of great significance is point 6. It indicates the possibility of the existence of not only elements that were as yet undiscovered, it also suggests that their physical and chemical properties could be defined as well. However, before we begin to discuss this problem in more detail, let us summarize the significance of Mendeleev's System. It should be noted that it included many elements which, together with the conception of periodicity, had already been taken into consideration in systems developed by other investigators. Mendeleev took into account all these characteristics simultaneously and devised his system more consistently than any other of the earlier 'classifiers', and—which is more important—having realized better than others the significance of such a system of the elements, he derived far-reaching conclusions concerning some physical and chemical properties of the elements and their compounds as yet to be discovered.

Mendeleev modified his System all his life, and as early as December 1870 he changed its arrangement from vertical to horizontal, distinguishing a period of seven typical elements—from lithium to fluoride—and five periods, each of which contained two rows with seven elements, respectively. Such a version of the table enabled the author to achieve a more accurate realization of the periodic changes of the properties of the elements and to formulate a few important

144   3. THE ELEMENTARY PARTICLE OF MATTER

conclusions. He presented them in an extensive work, already quoted 'The Periodic Regularity of Chemical Elements' published in the supplement to the journal *Annalen der Chemie und Pharmacie*; in 1871 Mendeleev defines the law of periodicity in the following way:

I define the law of periodicity as the mutual relations between the properties of the elements and their atomic weights which can be applied to all the elements. These relations have the form of a periodic function [106].

Next Mendeleev states that the arrangement of the elements according to their increasing atomic masses is more unequivocal than the earlier attempts at the classification of the elements according to their physical properties (distinction between metals and non-metals), chemical properties (valency, formation of acids or bases), and electrochemical properties. The results of the earlier methods frequently depended—in the opinion of the Russian scientist—on the conditions of measurement.

Mendeleev's predictions came true within the following dozen years. His success was also due to the fact that he predicted not only atomic weights but also some physical characteristics of the elements and their compounds unknown to him.

In 1875. Lecoq de Boisboudran, using a new spectroscopic method, discovered an element of atomic mass about 70 which he called gallium. Mendeleev realized that the properties of gallium correspond to those of an element of the same atomic mass which he had predicted and which should occupy the place in the same column in which aluminium was contained. He temporarily called this element eka-aluminium. Two other elements predicted by Mendeleev were discovered by the classical method of analysis. In 1879, Lars F. Nilson discovered eka-boron among the elements of rare earths. Later he renamed it scandium, and in 1886 Clemens A. Winkler discovered eka-silicon, also predicted by Mendeleev, and called it germanium. The following list (according to Karl Seubert) compares some properties of eka-silicon, predicted by Mendeleev, with those of germanium [117].

Apart from analogue of boron, silicon, and aluminium, in his later publications Mendeleev also predicted the existence of other elements,

Property	Eka-silicon (Es)	Germanium (Ge)
Atomic mass	72	72 (today 72.6)
Density	5.5	5.469
Atomic volume	13	13.2
Oxide	EsO_2	GeO_2
density	4.7	4.703
Chloride	$EsCl_4$	$GeCl_4$
melting point	below 100°C	86°C
density	1.9	1.887
Ethyl compound	$EsAe_4$	$Ge(C_2H_5)_4$
boiling point	160° C	160°C
density	0.96	a little lower than that of water

e.g. the element which is heavier by 80 atomic masses of hydrogen than the atom of tellurium, which he called dwitellurium, i.e. polonium, discovered in 1898 by Maria Skłodowska-Curie [118].

3.9. Energetism against the Corpuscular Theory

By the end of the 19th century atoms had already been distinguished from molecules, the former being arranged according to the law of periodicity. The kinetic theory of gases, then generally accepted, was based on the idea of corpuscularity. It included both the mutual attraction of remote particles and their repulsion when they approached one another too closely. In 1865, Joseph Loschmidt determined by physical methods the number of molecules contained in a definite volume of a gas under given conditions of pressure and temperature. The considerations then included, as we shall see in the next chapter, the spatial distribution of atoms forming a molecule and, furthermore, the complex texture of the atom itself began to be better understood. Nonetheless, many serious scientists and philosophers still doubted the actual existence of atoms and molecules and treated them merely as hypothetical particles which facilitated a convenient description of chemical phenomena. Reservations of that group of scientists found confirmation in a new branch of chemistry, which came to be known as physical chemistry, and precisely in one of its divisions—chemical thermodynamics.

Thermodynamics, or more precisely phenomenological thermodynamics, was developed in the mid-19th century by combining the concepts of dynamics and the science of heat. During the next few decades it was adopted by chemistry, and chemists eventually received the first quantitative theory which allowed them not only to account for many chemical processes but also to predict quantitatively the effect of phenomena that were as yet unknown. (In the preceding chapter we recognized the phlogiston theory as the first chemical theory, however its inferences were only qualitative.) The last 20 years of the 19th century marked a period of enormous success and popularity of the new theory. Its assertions and principles seemed to be absolutely unshakable. One of the tenets of the theory claimed that all phenomena which occur spontaneously in nature must be accompanied by an increase in the entropy of the Universe. Yet inferences which resulted from the atomic theory seemed to contradict this principle.

According to the assumptions of the corpuscular theory, particles in a gas are in a continuous, chaotic motion. It is conceivable, therefore, that at some particular moment they can all be concentrated in one half of the volume occupied by the gas and the other half will thus become empty. However, according to the laws of thermodynamics, such a phenomenon would have been accompanied by a decrease in entropy, so from the standpoint of thermodynamics it is absolutely impossible. In view of this obvious contradiction of inferences, thermodynamics rejected the idea of the real existence of atoms.

In the foreground of this group of scientists engaged in thermodynamics stood one of the creators of physical chemistry, the outstanding German scientist Wilhelm Ostwald. His and his adherents' views were shaped under the influence of the lectures of one of the creators of empiriocriticism, Ernst Mach, who held that in research on Nature we have only a knowledge of the relationships between phenomena. What we conceive of—according to Mach—changes easily, together with a change in our point of view. Ostwald did not

3.9. ENERGETISM AND CORPUSCULAR THEORY

recognize the real existence of the corpuscles of matter; therefore, he also rejected the conception of atomic and molecular weights. In reality, he asserted, there exists only energy, and in chemistry the amounts of particular elements should be determined by means of equivalent weights. Such reasoning was referred to as 'energetism'.

As late as 1904, after the discovery of radioactivity and its laws (see further), during a Faraday Lecture delivered in London Ostwald stated:

> It is possible, to deduce from the principles of chemical dynamics all the stoichiometrical laws (...). What we call matter is only a complex of energies which we find together in the same place. We are still perfectly free, if we like, to suppose either that the energy fills the space homogeneously, or in a periodic or grained way; the latter assumption would be a substitute for the atomic hypothesis. The decision between these possibilities is a purely experimental question. Evidently there exist a great number of facts—and I count the chemical facts among them—which can be completely described by a homogeneous or non-periodic distribution of energy in space. Whether there exist facts which cannot be described without the periodic assumption, I dare not decide for want of knowledge; only I am bound to say that I know of none [119].

This categorical assertion of Ostwald was denied in the following year by the progress of science. In 1905 Einstein proved that the two already known but still inexplicable phenomena could be accounted for only by adopting the periodic distribution of energy. One of them was a photoelectric phenomenon. Einstein explained it on the assumption that the electromagnetic radiation which produces it consists of portions of energy which he called 'photons'.

Another phenomenon, known from 1827, involved a continuous movement of suspended particles called the 'Brownian motion'. Its explanation was connected with the development of statistical thermodynamics, which evolved beside phenomenological thermodynamics. The main creators of this line of research were Ludwig Boltzmann, a professor at the University of Vienna and a Polish physicist, Marian Smoluchowski. Boltzmann, developing some of the views of Maxwell, introduced a statistical approach to thermodynamics. This approach allowed a reconciliation of the conclusions of thermodynamics with the ideas resulting from the corpuscular conception of the texture of matter. The basic contribution of Boltzmann was the introduction of the concept of probability. He went on to consider what was the probability that a given particle in an ensemble would have a definite energy; or, in other words, what energy the whole ensemble of particles would have. Boltzmann and Smoluchowski also considered how the particles of a certain ensemble could be distributed in space, i.e. what was the probability of finding a given particle at a particular point. This research led to the development of the theory of 'fluctuation'. The theory of fluctuation permitted to understand the fact that opposed processes which occur together simultaneously may not be equivalent. The course of events is not monotonic but undergoes accidental fluctuations. On this account, the state of the system also undergoes variable aberrations from equilibrium. The smaller the aberrations, the more probable they are. Thus, the concentration of all the particles of gas in one half of its initial volume is feasible, but this state would be such a great aberration from the equilibrium state that its probability is virtually zero.

Based on such statistical reasoning, in 1905 Einstein qualitatively explained the dependence of the value of the displacement of a macroscopic particle suspended in a liquid over a period of time. Such displacement, the result of Brownian motion, was explained by Einstein as a chaotic collision of the liquid molecules with this macroparticle. The theory of fluctuations was developed in 1904 by Smoluchowski. Employing this theory, independently of Einstein, in 1906 he derived a formula which explains Brownian motion. It differs from Einstein's formula only by a different value of the constant term. On the basis of his theory, Smoluchowski also calculated the probability with which different numbers of the particles of a suspension should appear in the field of view of a microscope. Observations of this kind were made by the Swedish chemist Theodor Svedberg. In his publication of 1907 entitled *Studien zur Lehre von den Kolloiden Lösungen* (Studies on Colloidal Solutions), he confirmed the full numerical conformity of his results with Smoluchowski's calculations. A review of this publication, written in 1908 by Ostwald, acknowledged the final victory of the corpuscular theory:

Elsewhere the reviewer voiced his opinion that actually the question, it seems, concerns the long-sought criterion of the kinetic theory and believes that he should not delay in asserting that the reason of rejecting this theory comes no longer in consideration since such a criterion has now been provided [120].

In the same year Jean Perrin, applying Laplace's hypsometric law to the distribution of the particles of gamboge in a vertical column, for the first time calculated the value of Avogadro's constant [121]. The results of this investigation are usually quoted from a more comprehensive monograph *Les atomes* published in 1913 by this author. A year after the first measurements of Perrin, in 1909, Stefan Dąbrowski who then collaborated with the French investigator, obtained a value of Avogadro's constant (that was in agreement with Perrin's results) on the basis of the hypsometric measurements of the suspension of mastic gum and on the basis of the investigations of Brownian motion [122].

Thus, it was not until the end of the first decade of the 20th century that the concept of the continuous texture of matter was finally rejected.

3.10. Stoichiometric and Non-stoichiometric Compounds

We have previously said that in the 19th century many chemists employed chemical equivalents rather than molar masses in their calculations. The former sufficed to confirm both the law of constant composition and the Daltonian law of multiple proportions. The latter law was gained additional importance due to the great significance of valency in the systematization of the elements in Mendeleev's Periodic System.

Dalton's law was—as we have already mentioned—basically incompatible with Berthollet's view that the elements can be combined in arbitrary proportions. As a matter of fact, as late as 1809 Gay-Lussac attempted to reconcile the two contradictory views arguing:

We must agree with M. Berthollet that chemical action occurs between molecules in an indefinite and continuous way irrespective of their number and proportion, and that generally we can obtain compounds of a very different proportion.

However, we must also agree that chemical action occurs more intensely when the elements remain in simple multiple relations or proportions [95, p. 23].

In the late 19th century, Berthollet's thesis was wholly rejected, whereas the law of multiple proportions gained general recognition. However, intensive research on gases and solutions was carried out in the latter half of the last century. It proved that molecules also interact, although these interactions are weaker than chemical actions. The concepts of hydrates and complex compounds were then developed. In 1891, Alfred Werner formulated a theory concerning these compounds (see the next chapter).

The above-mentioned research, and particularly the research on solid solutions, on the so-called metallic alloys and the alloys of organic substances, which was carried out in the early 20th century, proved that the law of multiple proportions is not a general law. In the years 1900–1914, the Russian physical chemist Nicolai S. Kurnakov, in his investigation of these alloys, applied a method of analysis of the dependence of various physicochemical properties on the content of particular constituents. He took into account the curves of alloy solidification, their thermal and electrical properties; on this basis he showed that the chemical entities that are formed in alloys have all the features of chemical compounds, except for a constant composition. For example, the alloys of thalium with bismuth in the range of 55–65 molar per cent of bismuth will melt at a definite temperature without a change in composition, just like pure substances. This temperature depends, however, on the alloy composition and has a maximum value for an alloy which contains 62.8 per cent of bismuth. Kurnakov proposed naming compounds of this type 'berthollides' in honour of Berthollet. Thermal analysis of alloys did not deny the fact that some elements combine with one another according to Dalton's law of multiple proportions. Kurnakov proposed calling such compounds 'daltonides' [23].

The problem of compounds whose composition was incompatible with Dalton's law was studied in the years 1931–1933 by the German chemist E. Zintl [124]. He showed that the elements of the first four main groups of the Periodic System always form compounds with the elements of the remaining three groups in accordance with Dalton's law; however, if both elements which form compounds belong to the first four groups, the composition of the compounds may be different. For example, the following compounds were found: three compounds of magnesium and zinc, their compositions being MgZn, $MgZn_2$, and Mg_2Zn_{11}; five compounds of sodium and lead: $Na_{15}Pb_4$, Na_5Pb_2, Na_2Pb, NaPb, and $NaPb_3$; seven compounds of sodium and mercury: Na_3Hg, Na_5Hg_2, Na_3Hg_2, NaHg, Na_7Hg_8, $NaHg_2$, $NaHg_4$. In the majority of these compounds the number of the constituent atoms of the particular elements is not compatible with their valency.

Such elements, referred to by some authors as 'Zintl's elements', thus form compounds in which the ratio of the elements cannot be expressed by small integers. At present daltonides and berthollides are more frequently called 'stoichiometric' and 'non-stoichiometric compounds'. The latter nowadays play a significant role in the production of semiconductors.

3.11. Parts of the Atom

In the late 19th century the opponents of the corpulscular texture of matter found confirmation of their reservations in thermodynamics. However, in the same period the first experimental evidence was obtained that showed that the atom is not the smallest quantity of matter. The first observation of this kind was made by Johann Wilhelm Hittorf in 1869. He noticed that in an evacuated glass tube equipped with a set of electrodes luminuous rays were given off from the negative electrode—the cathode—when a high voltage was applied to the electrodes. They were called 'cathode rays'. In 1876, William Crookes examined the properties of those rays and found that they propagate rectilinearly, and that objects laid in their way cast their shadows. The trajectory of the cathode rays was deflected, however, when the tube was placed in a magnetic field. This was proof that the particles of which those rays were composed have an electrical charge.

In 1886, Eugen Goldstein showed that another kind of rays with a opposite electrical charge escape from the tube through which cathode rays flow. Goldstein discovered their presence behind the cathode, when he bored a number of holes or 'canals' in it; those new rays were then called 'canal rays'. Thus there were two kinds of rays in the tube, and the particles which made up cathode rays must have had a charge opposite to those which made up the canal rays. In 1895, Jean Perrin, on the basis of the direction of the deflection of cathode rays in the magnetic field, proved that their particles carry a negative charge; he identified these particles with the carrier of an elementary negative charge which was called the 'electron' in 1891 by Johnstone Stoney. Therefore, the particles of canal rays must have had a positive charge.

In the years 1897 and 1898, Joseph John Thomson investigated the cathode rays, and Wilhelm Wien worked on the canal rays. On the basis of the magnitude of the deviation of the trajectory of both kinds of rays in the magnetic field, it was found that the particles which form canal rays are about 1840 times heavier than the electron. The magnitude of the deviation of the trajectory of electrically charged particles which move in the magnetic field depends on the ratio of the value of the charge carried by the particle to its mass. The value of this ratio in the case of particles which form cathode rays proved to be identical with that calculated for particles which cause electrically excited gases to emit visible radiation. These spectral calculations were based on the phenomenon of the splitting of spectral lines in the magnetic field (the Zeeman effect), and the theory of the electron developed by Hendrik Antoon Lorentz. On the other hand, the value of the ratio of charge to mass in the case of particles which form the canal rays was in agreement with that determined for ions carrying unit positive charge on the basis of Faraday's laws of electrolysis. These results proved that electrons and positive particles must exist inside the atoms.

The research on the cathode rays allowed the discovery of another phenomenon. In November 1895, Conrad Röntgen, a professor of physics at the University of Würzburg, stated that the fluorescing surface of glass on which cathode rays fall emits a new kind of rays which penetrate various solid bodies. In the next year, the French investigator Antoine Henri Becquerel, while attempting to determine the relationship between the fluorescence produced by different kinds

3.11. PARTS OF THE ATOM

of radiation and the emission of rays which penetrate solid bodies, discovered the spontaneous radiation of uranium. Maria Skłodowska-Curie and her husband Pierre Curie developed this line of research and separated two hitherto unknown elements from uranium ore: polonium and radium. They also found that these elements emit three kinds of rays of different permeability which behave in a different way in the magnetic field. These three kinds of rays were named after the first three Greek letters: alpha, beta, and gamma. 'Beta rays' were soon identified with the stream of electrons. In 1908, Ernest Rutherford, together with his collaborators, proved after arduous research that the particles of 'alpha rays' are ionized atoms of helium, and in 1912 Edgard Meyer showed that 'gamma rays' are an electromagnetic radiation whose wavelength is shorter than that of Röntgen's rays.

However, before the alpha particles were thoroughly investigated, Maria Skłodowska-Curie and other investigators had realized that the observed phenomena were not usual chemical reactions, but involved a transmutation of atoms themselves. Simultaneously, the views on the structure of the atom were changing. In 1904, J. J. Thomson proposed a hypothesis that electrons which—according to a previously proposed model—are inside a sphere consisting of positively charged particles are not arranged chaotically but occupy spaces in the form of rings or flat clouds equidistant from the centre of the sphere. Positive particles were still believed to be uniformly arranged within the spherical atom. This model was commonly referred to as a 'cake with raisins'.

However, this concept was denied by the experiments performed by Hans Geiger in Ernest Rutherford's laboratory on the scattering of alpha particles by thin metallic foils. In the year 1911, Rutherford concluded from the results obtained during these experiments that the atom contains a positive charge concentrated in a very small volume [125]. This part of the atom was called the 'nucleus' by Rutherford.

In the same year, Frederick Soddy found that, as a result of the emission of the alpha particle by an atom, another atom is formed, the element being shifted two places to the left in Mendeleev's Periodic System. Displacement of the elements in the Periodic System due to the emission of both alpha and beta particles by their atoms was observed in 1913 by four scientists, working independently. The Polish chemist Kazimierz Fajans was the first to present a full law of displacements at the session of the Chemical Society in Karlsruhe on 10 January. An account of this paper was published in *Chemiker Zeitung*, and a more extensive version was published in the journal *Physikalische Zeitung* on 15 February. In it the author pointed out the differences in the atomic masses of lead which came from two different radioactive series: thorium and uranium. An analogous law was published by A. S. Russel in the *Chemical News* on 30 January (this article contains a few errors, however). On 15 February, *Physikalische Zeitung*, besides K. Fajans' article, also published the work of G. Hevesy on the same subject. Finally, on 28 February, F. Soddy, referring to the publications of K. Fajans and A. S. Russel in the *Chemical News*, extended the ideas put forward by the Polish chemist. In the same year, Soddy also introduced the concept of isotopes, and the Dutch scientist Jan Atram van den Broek pointed out that the number of an element in Mendeleev's Periodic Table has the same value as the charge of its atomic nucleus.

Thus, it was finally confirmed early in the 20th century that atoms are complex structures, and—furthermore—it was found that some of them are spontaneously transmuted into atoms of other elements. The rate of this transmutation, however, depends on the kind of atom and for particular elements it may vary by an order of 20. The time after which half of the atoms of an element undergo transmutation varies between one thousandth of a second and thousands of millions years.

The idea of the chemists cherished silently for years—as Mendeleev described in 1871 the concept of the complex structure of the atom—eventually received a strong support and became well known. It was thus no wonder that the old hypothesis of Prout was revived in a new form. In 1918, A. Stewart and, two years later, William Draper Harkins propounded a view that the nucleus itself is composed of a certain number of positive nuclei of hydrogen atoms, which later became to be known as 'protons', and a small number of negative nuclear electrons. In a neutral atom the number of extranuclear electrons was believed to be equal to the difference between the number of protons and the number of nuclear electrons. This model gave rise to many reservations. The calculated value of the bond energy of nuclear electrons inside the nucleus was improbably high. Inconsistency was also found in another value, the idea underlying which will be briefly explained.

Considerations aimed at explaining the observed frequencies of lines in the atomic spectrum and the spectral changes when the emitting gaseous substances were placed in a magnetic field led to the conclusion that the value which characterizes the electron is—apart from its mass and charge—also its own angular momentum, commonly referred to as 'spin'. This hypothesis was proposed in 1925 by the Dutch physicists George Eugene Uhlenbeck and Samuel Abraham Goudsmit. It is worth noting that the first attempts at explaining the observed regularities in spectra by means of the photon angular momentum were made by the Polish physicist Wojciech Rubinowicz, who presented this concept in 1917. It was believed that protons must possess spin since electrons have it and, because angular moments are added vectorially, then particular atomic nuclei must also have characteristic spin values. The values of those spin were found to determine the distribution of intensities in the rotational spectrum of molecules. In 1930, Frank Rasetti investigated the intensity of rotational lines in the spectrum of light scattered by gaseous oxygen and nitrogen, but the result of the investigation did not confirm the spin value of the nuclei of these atoms calculated on the basis of the proton-electron model of the nucleus.

The new model of the nucleus was proposed in 1932 immediately after the discovery of new particles called 'neutrons'. The discovery of these particles took several years of experimental research. In 1919, Rutherford noticed that nitrogen, when bombarded by alpha particles, gives off protons and is transformed into an element whose atomic number is higher by one unit, i.e. oxygen. In the following year, Rutherford proposed an hypothesis in a public lecture that there may exist electrically neutral light particles which are a combination of the proton and the electron. Without being repelled by the nuclei they may collide with them and cause their transmutations more readily than the alpha particles [126]. Rutherford developed this idea in his lecture about the natural and artificial disintegration of elements which he delivered in Philadelphia in 1924.

The emission of particles (predicted by Rutherford) from nuclei was first observed by the two German physicists Walter Bothe and H. Becker in 1930 who did not realize the significance of the phenomenon. They found that when beryllium is bombarded with alpha particles, it emits a very penetrating radiation which they called beryllium radiation. In 1932, this discovery was confirmed by Irène and Frédéric Joliot-Curie who did not explain the phenomenon either. Frédéric Joliot-Curie admitted later that he failed then to find the right solution because he did not know of Rutherford's lecture from 1920. However, the pupil and co-worker of the English investigator, James Chadwick, was familiar with that lecture and in 1932 he showed that the beryllium rays observed by the German and French physicists are a flux of neutral particles—neutrons—predicted by Rutherford twelve years earlier. For this he was honoured by the Nobel Prize.

In the same year, Igor Tamm and Dimitri Ivanenko of the USSR, and Werner Heisenberg of Germany, working independently, proposed a hypothesis which is accepted to the present day. They postulated that the atomic nucleus consists of protons and neutrons. The number of protons is equal to that of the electrons orbiting outside the nucleus, and neutrons are particles whose mass is almost equal to that of the proton.

In 1933, Patrick Blackett and Giuseppo Occhialini discovered another particle after numerous experiments. Its mass was equal to that of an electron, and its charge differed from that of the electron only by its sign. This positive electron was named 'positron', and the negative one 'negatron'. In the same year, the investigators from three laboratories stated independently of one another that the positron is a very unstable particle because it will easily integrate with the negatron, yielding two quants of gamma radiation. This phenomenon was discovered by Irène and Frédéric Joliot-Curie and Carl David Anderson, as well as Lise Meitner and K. Philipp. The positron turned out to be the first discovery of a particle of an entirely new kind; each particle has an analogous equivalent. A pair of equivalent

3.11. PARTS OF THE ATOM

particles is characterized by equivalent masses, electrical charges of equivalent values but opposite signs, and different quantum numbers which determine particular properties that were unknown in 1933. Today such a pair is called a particle and its antiparticle.

In 1935, the Japanese physicist Hideki Yukawa was the first to conclude theoretically that protons, neutrons, negatrons, and positrons could not have been the only simplest particles of matter. He showed that the existence of particles whose mass is smaller than that of nucleons but greater than the mass of electrons is a necessary condition in order that the number of nucleons (protons and neutrons) should not be dispersed as a result of electrostatic repulsion. These particles are exchanged between nucleons which, thanks to the exchange, form a nucleus. Such a particle, having a mass about 270 times greater than that of an electron, as predicted by Yukawa, was discovered in cosmic rays by Cecil Frank Powell in 1947. This particle was then called 'meason pi' or 'pion'. Earlier, in 1937, Carl David Anderson and Seth Neddermeyer had discovered also another particle in cosmic rays, two hundred times heavier than the electron, which is today known as a 'muon'. At first it was erroneously identified with the Yukawa particle; later it turned out that its properties are rather similar to those of the electron than to the pion. Recent research has led to the discovery of several hundreds of kinds of particles, all of which are now called 'elementary particles'. Some of these particles can sometimes be included in the composition of the nucleus; most frequently they appear as products of reactions occurring between other elementary particles. Their masses range between zero and 21000 electron masses.

Particles belonging to that numerous set of elementary particles differ from each other not only by mass, charge, or spin. New fundamental properties of these particles have been discovered, such as isospin, hypercharge, and strangeness, which do not find any correspondence in the physics of macro-particles. We shall not explain the meaning of these concepts but it should be noted that—except for the mass—all the properties of elementary particles are quantized and may be described by appropriate quantum numbers, including the lepton and baryon numbers. These particles form pairs consisting of a particle and an antiparticle, similarly to the positron-negatron pair.

Collision of a particle with its antiparticle may cause a change in the corpuscular form of matter into its energetic form.

The existence of such a great number of elementary particles prompted scientists to consider that they cannot be the smallest particles of matter, but that even smaller particles may exist. In 1964 the American physicist Murray Gell-Mann made the first attempt at a systematization of the elementary particles, assuming that particular particles are different combinations of several constituents. He coined a facetious name for these constituents—'quarks' (small goblins)—derived from James Joyce's book 'Finnegan's Wake'. The charges and baryon numbers of Gells-Mann's quarks are multiples of one third of the unit values of those quantities. The quark hypothesis is now recognized by the majority of investigators. The existence of six kinds of quarks allows us to make a rational systematization of the known elementary particles. However, the quark hypothesis required the introduction of some new elementary properties which would distinguish them. These properties have absolutely no correspondence to the macroscopic world; therefore they have been denoted by fanciful conventional names of 'colour', 'strangeness', and 'charm'. All these properties are quantized.

Until now no one has managed to separate a quark from any elementary particle. What is more, many physicists believe that quarks interact so strongly that they do not occur as isolated particles; they may exist only in combination with other quarks.

And thus, after twenty-five centuries, like the Greek philosophers, we have again come to consider what are the basic properties that distinguish the most elementary particles. The Greek philosophers sought these basic properties in the observations of everyday phenomena; the present-day physicists must refer the new basic properties to the models they have created themselves.

For centuries attempts have been made to separate the carriers of these basic properties in order to obtain them in the pure state. Today physicists have abandoned such a project; single quarks cannot be investigated experimentally. For many decades, the element was recognized as the limit of chemical analysis; it thus represented the limit of the chemists' skills. Today what is identified as the smallest particle of matter reflects the current limit of our cognition.

CHAPTER 4

The Structure of Chemical Compounds

4.1. Structure in Inorganic Chemistry

It was not until the second half of the last century that the structure of chemical compounds, i.e. the molecular structure of different substances, became an object of interest. After the congress in Karlsruhe the concepts of the atom and the molecule were finally discriminated. Naturalists have always created concepts by which they attempted to explain the observed phenomena. As we have said earlier, it was already assumed in ancient times that there exist unobservable atoms by which philosophers tried to explain the observed phenomena. The models of those atoms were macroscopic bodies whose shapes and surfaces served to explain their properties. The atoms of Democritus had definite spatial shapes, and Plato's atoms of the four basic elements were, as we have noted, geometrical regular solid bodies whose walls were built of equilateral triangles. Plato thought that each of those solids-atoms could be decomposed to triangles, and from the triangles new solids-atoms could be built. So an icosahedral atom of water could be transformed into two octahedral atoms of air and one tetrahedral atom of fire.

Also Plato's pupil, Aristotle writes about the shape of the elements, although he understands the role of shape in a slightly different way:

... it is manifest that the simple bodies are often given a shape by the place in which they are included, particularly water and air. In such a case the shape of the element cannot, persist, for, if it did, the contained mass would not be in continuous contact with the containing body; while if its shape is changed, it will cease to be water, since the distinctive quality is shape [23, 306b].

Thus, according to Aristotle, the geometric configuration of atoms depends on the surroundings; however their shape is not the shape of Plato's atoms, but only the shapes or combinations of shapes which fill space continuously.

4. THE STRUCTURE OF CHEMICAL COMPOUNDS

For the whole period of Greek science, naturalists were not so much interested in the structure of substances as in their ability to interact. This problem will be dealt with in the next chapter. The first traces of interest in the structure of substance can be found in Lavoisier, although earlier Boyle seems to have realized its role. In 'The Sceptical Chymist', he writes:

(...) these differing substances that are called elements or principles (...) are only various schemes of matter or substances that differ from each other, but in consistence (...) and some very few other accidents, as taste, or smell, or inflamability, or the want of them. So that by a change of texture (...) the same parcel of matter may acquire or lose such accidents as may suffice to denominate it salt, or sulphur, or earth [32, p. 199].

How great was the significance of the structure of matter for Boyle, we can learn from the following fragment:

For I am apt to think, that men will never be able to explain the phaenomena of nature, while they endavour to deduce them only from the presence and proportion of such or such material ingredients, and consider such ingredients or elements as bodies in a state of rest; whereas indeed the greatest part of the affections of matter, and consequently of phaenomena of nature, seems to depend upon the motion and the contrivance of the small parts of bodies [32, p. 178].

Thus Boyle was one of the first scientists to realize the significance of the structure of bodies. This is made explicit in Carneades's statement:

But how colours do, nay, how they may, arise from either of these principles, I think you will scarce say that any has yet intelligibly explicated. And if Mr. Boyle will allow me to shew you the experiments which he has collected about colours, you will, I doubt not, confess that bodies exhibit colours, not upon the account of the predominancy of this or that principle in them, but upon that of their texture, and especially the disposition of their superficial parts, whereby the light rebounding thence to the eye is so modified, as by differing impressions variously to affect the organs of sight [32, p. 176].

Therefore, in the period when the qualitative approach dominated, scientists were concerned with what we have called in the preceding chapters the texture of matter.

Along with the development of quantitative methods, the problem

4.1. STRUCTURE IN INORGANIC CHEMISTRY

of structure gained increasing importance, which in turn involved the problem of terminology.

Due to the discovery throughout the 18th century of many new substances and new techniques of obtaining them, it became increasingly troublesome to call them after the names of their discoverers or places of discovery. First I. P. Macquer and A. Baumé began to name metallic salts by the names of acids and metals from which these salts could be obtained. This idea was then extended by P. Boucquet and A. de Fourcroy, and finally given a concrete form in 1782 by L. B. Guyton de Morveau, who formulated in the May issue of the journal *Observation sur la physique* following principles for creating chemical names:

The first principle: Proposition is not a name: chemical entities and products should have names which denote them in all circumstances without the need of circumlocutions (...), The second principle: Names should be as much as possible adjusted to the essence of things.

It is a mistake, de Morveau argued, to call vitriolic acid (sulphuric acid—R.M.) oil of vitriol because this compound has nothing to do with oils.

Conclusions: Firstly, it would be best for a proper noun to denote the simplest entity (...), and the expression that modifies or qualifies it should be added adjunct (...). Secondly: The name of a chemical compound is clear and precise only when it reminds us of its constituents by names that refer to their nature (...). Thirdly: The names of discoverers which cannot be adjusted to things individually or generically should be forbidden in terminology (...). The third principle: If we do not know sufficiently any quality that might be a basis for naming, a name that expresses nothing is better than that which might express a wrong concept (...). The fourth principle: Choosing new names, one should prefer those which have roots in the most universal dead languages, so that a word could be easily identified by its sense and sense by a word (...). The fifth principle: Names should be carefully chosen according to the spirit of the language for which they are created [127].

Having developed the above principles, A. L. Lavoisier, L. B. Guyton de Morveau, L. C. Berthollet, and A. de Fourcroy, during four sessions of the Royal Academie des Sciences from 18 April to 13 June 1787, presented the principles for creating the names of inorganic compounds and proposals for new names in the French language [128]. Two years later, A. L. Lavoisier decided to present these principles to a wider public in his textbook *Traité élémentaire de chimie*. The scope of his book exceeded his initial aim because—as he wrote in the introduction—"it impossible to improve the language without improving science, or science without improving the language, and whatever facts or ideas are born, they will be given false impressions, if we are not able render them by exact expressions" [36, p. I].

The principles of creating chemical names proposed by the French scientists were soon adopted all over the world, and they have remained a basis of chemical terminology to the present day. Therefore, it was of great importance that Lavoisier specified for the first time which substances should be regarded as simple bodies, i.e. elements, and which substances are compounds or mixtures.

Hence Lavoisier's considerations concerning acids and bases may be regarded as the first modern attempt at determining the structure of compounds. As we have already mentioned, the second group of elementary bodies in *Traité élémentaire de chimie* contained those elements whose oxides had acidic properties, while the third group contained metals, i.e. elements whose simple oxides were basic oxides. We remember that Lavoisier predicted (sometimes correctly, sometimes wrongly) that a given element can form several oxides of different oxygen content, and the more oxygen in such a compound, the stronger acidic properties it might have. Some oxides of such metals as arsenic or iron, which contained much oxygen, were supposed to be acids.

As we know from Chapter 2 'oxides at a higher degree of oxidation' were referred to by Lavoisier as acids, although according to the present-day notions they were only oxides. On the other hand, compounds which we now call acids and bases were for Lavoisier solutions of oxides in water; he did not believe that a chemical reaction could have occurred between oxides and water; he treated water only as a solvent. It should be remembered that this nomenclature was used for many decades, even when it was already known that oxygen acids were chemical combinations of oxides with water. For example, in his work entitled 'Polish Chemical Terminology' published in 1853, Czyrniański refers to oxides as acids, and their combination with water

4.1. STRUCTURE IN INORGANIC CHEMISTRY

as monohydrous acids; thus $SO_3 + HO$—was called hydrous sulphate or monohydrous sulphuric acid [129].

Since Lavoisier believed that all acids must contain oxygen, and salts were, he thought, combinations of salinizing (acidic) principles with salinizable bases which also contained oxygen; therefore salts must have been oxygen combinations, too. For this reason, in the early 19th century, when chlorine was found to be an element and not a combination of 'muriatic radical' with oxygen, chlorides, including table salt, were not regarded in chemistry as salts. It was not until 1827 that J. J. Berzelius recognized them as 'halogen salts'.

The Daltonian symbols for complex atoms discussed in the preceding chapter cannot be treated as attempts at explaining the structure of molecules, either; their objective was only to illustrate the proportions in which simpler constituents form a more complex compound. However, the first theories of the structure of compounds were set forth on the basis of the discovery of chemical phenomena connected with electricity. Those theories also concerned the ability of particular substances to react; this problem will be discussed in the next chapter.

The Italian physician Luigi Galvani was the first to deal with an electric cell, although he was not in fact aware of it. He noticed that an electric current passed through a dead frog's leg when it was touched by hooks made of two dissimilar metals. The mechanism of 'animal electricity', as Galvani called it, was accounted for by Alessandro Volta. On the basis of his considerations concerning Galvani's experiment, he constructed an electrical pile in 1800 in which a number of discs made of two kinds of metal were placed alternately, being separated by porous discs soaked in brine. Some discs were made of copper, brass, or silver, and the others of tin or zinc.

Electricity had already been used for the electrochemical decomposition of water into hydrogen and oxygen. In 1789, the Dutch chemists Paets van Troostvijk and Jan Rudolf Deiman succeded in this operation, using the electrostatic machine as the source of electric potential [130]. In 1800, this result was obtained by W. Nicholson and A. Carlisle applying the Voltaic pile. It was repeated in 1806 by Humphry Davy, who in the next year isolated potassium and sodium by the same method. He interpreted the observed phenomena in the following way: on approachning one another, the two substances became charged with opposite kinds of electricity. Although the opposite kinds of electricity are mutually neutralized when a compound is formed, after applying an electrical voltage to the electrodes, parts of this compound are separated and tend to travel towards the respective electrodes. Davy believed that chemical phenomena are closely interrelated with electrical phenomena.

On the basis of the above discoveries Berzelius published, in Swedish in 1818 and in French in 1819 an electrochemical theory of the structure of chemical compounds. This theory contributed significantly to the development of chemistry. Berzelius wrote:

We now believe that we know with certainty that bodies which are ready to be combined exhibit free opposite electricities which are intensified when they approach from the temperature at which a compound is formed to that at which during compounding the two opposite electricities disappear with a considerable rise of temperature, frequently so great that the fire explodes. On the other hand, we have the same certainty that compound bodies exposed to (...) the action of electrical fluid caused by the discharge of the pile, are separated and regain their original chemical and electrical properties at the same time when the electricities which act on them disappear.

On the grounds presented above, Berzelius formulates the final conclusion:

(...) in all chemical compounds there occurs neutralization of opposite electricities, and this neutralization produces fire in the same way as it does in the discharges of an electrical bottle, electrical pile, and thunder (whereas) the latter three phenomena are not accompanied by a chemical bond [131].

Berzelius noted, however, that the same compound as well as the same simple body can exhibit electropositive properties in relation to some bodies and electronegative properties in relation to others; it is thus necessary to consider the properties of one constituent with respect to another. Further, he held that oxygen is the most electronegative element, i.e. all other elements are electropositive with respect to oxygen. On this basis he listed the following electrochemical series in which every element is electropositive with respect to the elements that precede it, but electronegative with respect to the elements that follow it: O, S, N, Cl, F, Se, Mo, Cr, W, B, C, Sb, Tl, Tm, Si, Os, Au,

4.1. STRUCTURE IN INORGANIC CHEMISTRY

Ir, Ro, Pt, Pd, Hg, Ag, Cu, Ni, Co, Bi, Sn, Zr, Pb, Ce, U, Fe, Cd, Zn, Mn, Al, Y, Be, Mg, Ba, Na, K.

In the text book published in 1827, Berzelius describes more precisely the mechanism of the formation of compounds:

> If these electrochemical views are correct, then it follows that all chemical compounds depend exclusively on two opposing forces: positive and negative electricity; therefore, either should be composed of two parts combined as a result of an electrochemical reaction, assuming that there is no third force. Hence, it follows that every compound (...) can be divided into two parts, one of which being electrically positive, the other negative [132].

In Berzelius' statement we can see a trace of the Greek conception of contrarieties. His theory was called a 'dualistic electrochemical theory of the structure of chemical compounds'.

However, it follows from the above quoted fragments that Berzelius believed that isolated atoms or groups of atoms are electrically neutral, and they begin to exhibit electrical properties only when they interact with one another. Electrical charge is not therefore their specific quality, the more so as the same atom (except for an oxygen atom) or a group of atoms can assume a positive or negative charge depending on their properties with respect to another group of atoms. The electrical properties of a group of atoms depend, Berzelius held, on the electrical properties of constituent atoms. If either of the two different atoms, e.g. of sulphur and potassium, combines with the same third atom, e.g. of oxygen, these combinations (sulphur oxide and potassium oxide)—according to Berzelius—will have the same electrical properties with respect to one another as the primary atoms, i.e. the atoms of sulphur and potassium. Sulphur oxide (acid) will thus be electronegative with respect to electropositive potassium oxide (base), and both oxides can combine into potassium sulphate (salt). Thus electronegativity was believed to correspond to acidic properties, and electropositivity to basic properties.

However, Berzelius' electrochemical theory requires some comment. Firstly, it seemed to suggest that only atoms or groups of atoms which induce opposite electric charges on approaching one another would be combined. Thus, they had to be different atoms because no explanation be could have been given at that time why one of the two identical atoms became electrically positive and the other electrically negative. Therefore, this aspect of Berzelius' theory gave a significant support to the consequence of Dalton's view that two identical atoms cannot form a compund atom and hence cannot combine into one molecule.

Secondly, a problem arises what electrical charge is actually attributed to atoms and groups of atoms. According to Berzelius, no atom constantly contains the same amount of electricity. This amount depends on the conditions to which the atom is subjected. A given element seems to have some aptitude to receive a charge after interaction with another element, which was in those days referred to as affinity, although as a matter of fact it was similar to the present-day concept of electronegativity introduced by Linus Pauling in the 1940s.

However, the 19th-century investigators held that an electrical charge is one of the distinguishing characteristics of particular atoms. Such an approach was justified by the development of electrochemistry, and particularly electrolysis. It followed from the electrolysis of water carried out by Christian Grotthus in 1805 that each atom and each group of atoms carries a specific amount of electricity. In 1815, Davy believed that potassium sulphate consists of positive potassium and negative sulphur oxide, and that it is not a combination of two oxides, as Berzelius had stated. Finally, the laws of electrolysis formulated by Michael Faraday in 1833 also confirmed that definite amounts of positive or negative electricity are permanently connected with atoms and their groups. Moreover, as almost fifty years later H. Helmholtz [133] pointed out in his Faraday Lecture, those quantities of electricity which characterize the particular ions did not depend in the least on their affinity for one another, as it might have been suspected from Berzelius' theory. The term *ion*, which was introduced by Faraday, has been used until today. It is derived from a Greek word which means 'to wander'; the prefixes *ana* and *kata* (in the terms anion, cation, anode, cathode), which in Greek mean up and down, denote opposing directions; i.e. the 'wandering' of ions with opposite signs to the electrodes. In view of the development of electrochemical research, the 19th-century chemists found it hard to understand that the same atom can exhibit different electrical properties in relation to different atoms. In spite of reservations the dualistic theory was still

4.1. STRUCTURE IN INORGANIC CHEMISTRY

upheld, perhaps due to the authority of Berzelius, until his death in 1848.

Berzelius' authority was not shaken by the fact that he himself had to reject some of the inferences arising from his theory. Namely, it followed from his theory that the definite numbers of particular atoms can form only one compound. (Let us note that in the first half of the 19th century it was still not possible to determine the absolute number of atoms making up a molecule but only their proportions.) In spite of this conclusion, however, it was shown experimentally that there exist pairs of compounds characterized by entirely different properties in which the relative contents of particular atoms are identical. In 1823, Liebig, then a collaborator of Gay-Lussac found that the exploding silver fulminate—according to the present-day symbolism CNOAg—has the same composition as the silver salt of cyanic acid discovered a year earlier by Friedrich Wöhler—NCOAg. In 1827, Faraday published information that ethylene (which he called olefiant gas) contains atoms of carbon and hydrogen in the same proportion as another gas which is now called butene. In 1830, Berzelius himself discovered that tartaric and racemic acids contain the same atoms in equal proportion, but nonetheless they are different from each other in respect of their optical properties: tartaric acid rotates the plane of polarization of light, and racemic acid does not exhibit such a property. Berzelius also proved the existence of tin oxides of identical composition but different properties. In connection with the discoveries of Liebig, Wöhler, Faraday, and his own in 1832, he admits:

In physical chemistry it was long taken as axiomatic that bodies of the same composition, having the same constituents in the same proportions, necessarily must also have the same chemical properties.

The investigations of Faraday appear to indicate that there may be an exception to this if two similarly composed bodies differ in that the composition of one contains twice as many elementary atoms as occur in the other, although the proportions between the elements remain the same. It is thus with the two gaseous hydrocarbons, olefiant gas, which is CH, and the other more compressible gas described by Faraday (butylene) which is C_2H_4 and accordingly has twice as great a specific gravity as the former, the similarity in composition is

168   4. THE STRUCTURE OF CHEMICAL COMPOUNDS

only apparent, for the compound atoms are still definitely different, the relative numbers of elementary atoms being equal but the absolute numbers unequal.

Recent researches have now shown that the absolute as well as the relative numbers of elementary atoms may be the same, their combination taking place in such a dissimilar way that the properties of equally composed bodies may be different [134, p. 44].

He concludes his article:

Since it is necessary for specific ideas to have definite and consequently as far as possible selected terms, I have proposed to call substances of similar composition and dissimilar properties isomeric, from the Greek *iso meres* (composed of equal parts) [134, p. 47].

Thus, Berzelius was the originator of the concept of isomerism which contradicted his own theory of the structure of chemical compounds.

4.2. The First Theories of the Structure of Organic Compounds

In the first decade of the last century other facts were also discovered, which could not be reconciled with an electrochemical dualistic theory. We have mentioned in Chapter 1 that it was a period of the development of the chemistry of carbon compounds. Previously, eighteenth-century and earlier textbooks had made a distinction between the kingdom of minerals, plants, and animals. In the 18th century scientists wrote about structures 'organized' by 'vital force' (*vis vitalis*). Some of the carbon compounds, e.g. alcohols, had been extracted as early as the Middle Ages by distillation, although anhydrous alcohol was obtained for the first time by T. E. Lowitz in 1796. The 'organized' substances were investigated by combustion. Early in the 19th century, 'organized' substances came to be known as 'organic', and application of mild oxidizing and reducing agents enabled many new compounds to be produced. The term 'ladder of burning' came to be used at that time. As a result of the first methods of synthesis a few years late a 'ladder of synthesis' was introduced. In 1783, Scheele isolated hydrogen cyanide and potassium cyanide by heating cyanide dyes with acetic acid. In 1787, Berthollet replaced an atom of hydrogen by an atom of chlorine in hydrogen cyanide by treating it with gaseous

4.2. THE STRUCTURE OF ORGANIC COMPOUNDS

chlorine, thus chlorination was the first known substitution reaction. By chlorination of ethylene, Jan Deiman obtained dichloroethane in 1795. In 1811, John Davy synthesized phosgene, and in 1831 Justus Liebig obtained chloral and chloroform. In 1839, Henri Regnault performed the further chlorination of chloroform to obtain carbon tetrachloride. In 1825, Faraday established the molecular formula of benzene liberated from the deposits of illuminating gas obtained by heating whale and cod-liver oil. In the 1840s pure benzene (then generally called benzol) was for the first time obtained from the products of distillation of coal tar, although this technique had been described by Glauber as early as 1649.

The next important developments included the synthesis of urea made by Wöhler in 1828 (this will be discussed in more detail in the next chapter), as well as nitration, sulphonation, and chlorination reactions of benzene discovered in 1834 by Eilhardt Mitscherlich who also developed molecular (not structural!) formulas for the products obtained [135].

The composition of nitrobenzene proposed by Mitscherlich provoked Berzelius' objection in the following year. The latter believed that a compound which contained so much oxygen and nitrogen should have electronegative, i.e. acidic, properties and it should interact strongly with electropositive bases; yet nitrobenzene does not react with bases. Thus, this great number of new compounds, which could hardly be squared with the dualistic theory of Berzelius, required the formulation of new theories which would systematize compounds, particularly organic compounds. During the course of twenty years (1837-1856) four new theories were developed. They treated molecules as a whole and not as a structure made up of two interacting parts. They included Dumas' and Liebig's theory of radicals of 1837, Dumas' theory of substitution and types of 1840; in 1854 Laurent presented a theory of nuclei (kernels), and in the years 1853-1856 Gerhardt proposed a unitary system which was a combination of the theory of types and nuclei.

In 1837 Dumas and Liebig in *Note sur l'état actuel de la chimie organique* wrote:

4. THE STRUCTURE OF CHEMICAL COMPOUNDS

Chemist have recognized that in mineral substances there exist bodies which act as elements (...). Of course, that which they call element, or undecomposable substance has been considered as such only with regard to the state of acquired experience (...). (...) organic chemistry possesses its own elements which at one time play the role of belonging to chlorine or to oxygen in mineral chemistry and at the same time, on the contrary, play the role of metal. Cyanogen, amide, benzoyl, the radicals of ammonia, the fatty substances, the alcohols and analogous compounds—these are the true elements on which organic chemistry is founded and not at all the final elements, carbon, hydrogen, oxygen and nitrogen—elements which appear only when all trace of organic origin has disappeared.

These elements of the structure of compounds are referred to by the authors as radicals. Next they draw the following conclusion:

In mineral chemistry the radicals are simple; in organic chemistry the radicals are compound; that is all the difference. The laws of combination and of reaction are otherwise the same in these two branches of chemistry [136].

However, these organic radicals could not be isolated in the pure state, which caused other investigators to raise objections to the new theory. This theory could still be reconciled with the dualistic theory. In the following year Liebig proposed a new definition of acid to replace Lavoisier's theory of acid as a compound containing an oxygenic principle, i.e. oxygen. According to Liebig, an acid is a compound in which hydrogen can be replaced by a more electropositive metal [137]. It followed from Liebig's theory that there might be acids which contain several hydrogen atoms that could gradually be replaced by metal atoms. Berzelius could not agree with such a conclusion and the publication of this theory by Liebig even caused an estrangement between the two investigators.

In 1839, Dumas obtained trichloroacetic acid by chlorination of acetic acid. Thus he showed that in organic compounds hydrogen can be replaced not only by metals, regarded as more electropositive than hydrogen, but also by the more electronegative chlorine [138]. He also showed that both acids have similar chemical properties because they are both decomposed, giving off carbonic acid (CO_2). In the process methane is liberated from acetic acid, and chloroform from trichloroacetic acid. This observation prompted him to refute the dualistic

4.2. THE STRUCTURE OF ORGANIC COMPOUNDS

theories and to replace them by the theory of types. He makes a more precise account of it in a work published in the following year.

Lavoisier distinguishes in each of them (i.e. molecules—R.M.) a deflagrating and deflagrated element; electrochemical theory sees a negative body in the latter, which in fact refers to the same idea.

On the other hand, if we consider chemical compounds as being made up of particles, like planetary systems, supported by different molecular forces whose resultant creates affinity, we do not find it necessary to apply widely the law of dualism assumed by Lavoisier [139].

Although Dumas, in a work of 1839, states that both the above-mentioned acids belong to the same chemical type, yet it was not until 1840 that he provided a comprehensive definition of the concept of type:

By the bodies of the same chemical type we understand thus such bodies which exhibit the same principal reactions and are manifested only in bodies composed of the same number of equivalents, in which we may assume that their equivalents are interconnected in the same way [139].

However, defining his theory, Berzelius asserted that compounds of different densities, odours, and boiling temperatures, i.e. compounds with different physical properties, e.g. both acids according to Dumas, cannot be recognized as similar. On the other hand, in accordance with his theory, he presented these acids not as single compounds but as combinations of two compounds, i.e. paired compounds. Thus, according to Berzelius, acetic acid was a compound of the formula: $C_2H_3 + C_2O_3 \cdot HO$, and trichloroacetic acid was a compound of the formula: $C_2Cl_3 + C_2O_3 \cdot HO$.

As we can see there were several conflicting theories of the structure of chemical compounds in the 1840s. As has been noted, as a rule no distinction was made between chemical equivalents and molar weights (mass) of particular compounds, and the term 'atom' was frequently applied to what we now call 'atom', 'molecule', and sometimes 'equivalent' as well. Besides, only the ratios of elements in a compound and not absolute numbers of atoms in a molecule were known from experiments. For example, as late as 1854 Laurent considered whether the formula of acetic acid should be presented as CH^2O, $C^4H^8O^4$ or $C^2H^4O^2$ [90, p. 7].

4. THE STRUCTURE OF CHEMICAL COMPOUNDS

The dualistic view of the structure of compounds (including organic compounds) prevailed until the end of the first half of the 19th century. However, Laurent and Gerhardt tried to persuade chemists that a compound should be treated as an entity. Therefore, Gerhardt called his conception a unitary theory.

From 1836 Laurent investigated the influence of the substitution of a hydrogen atom in the particular series of carbon compounds on their chemical properties. He called a 'series' a group of compounds of the same number of carbon atoms analogously combined with one another, and within a series particular hydrogen atoms could be replaced by another single atom (e.g. a chlorine atom) or by a group of atoms [90, p. 363].

Any group of the series had its 'nucleus' (French *noyau*) in which hydrogen atoms could be replaced by other single atoms (see below). Thus Laurent writes:

I have attempted to find out if there is something analogous to the parent cell in all parts of the same chemical tree; briefly speaking, if there is a nucleus common to all the compounds in this series, a nucleus that would allow us to understand why these compounds can be mutually transformed into one another. While considering the crystalline form, analogy, reaction, and mainly stabilities of halydes and naphthalene (...) I came to the following conclusions: 1° hydrocarbons and their halydes form groups and stable nuclei analogously, when we consider the number and arrangement of their atoms but they are changeable when we consider the nature of these atoms, which can be the atoms of hydrogen, chlorine, bromine, nitrogen peroxide as well as atoms of other bodies which occupy the place of hydrogen; 2° perhalydes corresponding to aldehydes contain hydrocarbon or chloride which forms the nucleus or a separate group. Briefly, I conclude that in the case of the compounds of eterine or dihydrocarbon (previous names of ethylene—R.M.) the following system exists:

halydes	perhalydes	aldehydes	acids
	$C^2H^4.H^2$		
C^2H^4	$C^2H^4.Cl^2$	$C^2H^4.O$	$C^2H^4.O^2$
C^2H^3Cl	$C^2H^3Cl.Cl^2$	$C^2H^3Cl.O$	$C^2H^3Cl.O^2$
$C^2H^2Cl^2$	$CH^2Cl^2.Cl^2$	$C^2H^2Cl^2.O$	$C^2H^2Cl^2.O^3$
C^2HCl^3	$C^2HCl^3.Cl^2$	$C^2HCl^3.O$	$C^2HCl^2.O^2$
C^2Cl^4	$C^2Cl^4.Cl^2$	$C^2Cl^4.O$	$C^2Cl^4.O^2$

[90, pp. 399–401].

4.2. THE STRUCTURE OF ORGANIC COMPOUNDS

The basic nucleus here is ethane devoid of two hydrogen atoms (in the chloride group ethane is identified with methane). In that nucleus Laurent replaced the hydrogen atoms successively by the chlorine atoms and obtained different chlorinated compounds; adding other groups (Cl_2, O, O_2) gave a whole gamut of compounds in the series considered.

Simultaneously, and in consultation with Laurent, Dumas' collaborator Gerhardt developed the unitary theory. Apart from the above theory of radicals propounded by Dumas, and Laurent's theory of nuclei, he also considered the conception of 'homologous series' and the 'conception of types'. In 1840, Dumas and Jean Stas arranged aliphatic alcohols in a series, and in 1844 the former constructed an analogous series from the fatty accids. On this basis, in his textbook *Précis de chimie organique* published in 1844, Gerhardt stated that the reactions of compounds which differ only by the content of CH_2 groups, i.e. those which belong to a series which he called 'homological', can be described by the same equations, and these compounds have similar chemical properties.

The idea of types was quite new, too. In 1850, August Hofmann extended Dumas' conception of types. He considered amines as compounds of the ammonia type; i.e. he believed that the molecules of amines are actually the molecules of ammonia in which the hydrogen atoms were successively replaced by the hydrocarbon groups. In the following year, Williamson, in a paper on etherification, while justifying the correctness of a formula H—O—H for water, pointed out that:

Alcohol is (...) water in which half of the hydrogen is replaced by carburetted hydrogen and aether is water in which both atoms of hydrogen are replaced by carburetted hydrogen: thus

$$\begin{matrix} H \\ \end{matrix}\!\!\!\!\searrow\!\!\!\!\begin{matrix} \\ O \\ \end{matrix} \qquad \begin{matrix} C^2H^5 \\ \end{matrix}\!\!\!\!\searrow\!\!\!\!\begin{matrix} \\ O \\ \end{matrix} \qquad \begin{matrix} C^2H^5 \\ \end{matrix}\!\!\!\!\searrow\!\!\!\!\begin{matrix} \\ O \\ \end{matrix} \qquad [140].$$

Gerhardt explained the unitary theory in his textbook *Traité de chimie organique* whose four volumes appeared in the years 1853–1856. According to this theory, organic compounds are combinations of two radicals or nucleii, which Gerhardt called 'residues'. In the Introduction to his *Traité* he writes:

4. THE STRUCTURE OF CHEMICAL COMPOUNDS

Arranging organic compounds in series, i.e. establishing the laws according to which properties in a given type change by substituting one element or a group of elements for another element, is always the aim of a chemical philosopher. In the (present) state of science, organic compounds can be arranged in three or four types, each of them being able to form a series (...), these types are:

water H^2O hydrochloride acid HCl
hydrogen H^2 ammonia H^3N.

By exchange of their hydrogens into certain groups these types produce acids, alcohols, ethers, hydrides (hydrocarbons—R.M.), organic chlorides, acetones (ketones—R.M.), and alkalies [140].

According to Gerhardt, alcohol or ether can be obtained from the water type by substituting—as Williamson had already suggested [141] —one or both hydrogen atoms for —CH_3 or —C_2H_5 groups. Moreover, if hydrogen is substituted for groups containing oxygen atoms, besides the carbon and hydrogen atoms, then acids, esters, and acid anhydrides will be formed. Gerhardt expressed these compounds by the following formulae:

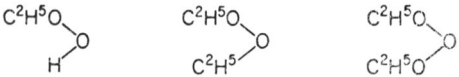

Gerhardt had not by than distinguished the oxygen atom which was attached to the carbon atom by a single bond, forming a kind of bridge between this atom and any other atom, from an oxygen atom of the carbonyl group; the concept of the double bond did not yet exist in his lifetime. Therefore, he treated the $CH_3(CH_2)$—and $CH_3(CO)$—groups all in a similar manner.

Substitution reactions of hydrogen atoms in the hydrogen type, analogously to those discussed above, lead, according to Gerhardt, to the obtaining of hydrocarbons and aldehydes, whereas

(...) the type of hydrochloric acid gives on one hand hydrochloric ethers (...), and on the other (...) monobasic acids, such as acetyl chloride or benzoyl chloride (...). Finally the type of ammonia produces alkalis which can combine with acids (amines—R.M.), or amides which can combine with bases (oxides of silver, mercury, copper, etc.) depending on whether substitution of ammonia hydrogens has been made through organic acid (radicals—R.M.) [140, vol. I, Introduction].

Summing up his considerations, Gerhardt writes:

I call a unitary method all the principles which I apply to chemical investigations and which are based on the choice of the unity of a molecule and the unity of reactions for comparison of chemical interactions of bodies [140, vol. IV, p. 585].

Gerhardt points out that the unitarian treatment let us represent all properties of molecule whereas "dualistic fofrmulae express only one or two relations and they never give the true image of molecular constitution" [140, Vol. II, p. 562].

Developing Gerhardt's conception Odling and Kekulé have introduced new types, e.g. type H—S—H and type CH_4.

4.3. Valency

The attempts to systematize compounds that we have discussed above, as well as the research on the ratio of the numbers of particular atoms (equivalents) contained in a given molecule (or a compound), led to the gradual development of the concept of valency. The actual term 'valency' (German *Wertigkeit*) was introduced as late as 1868 by K. Wichelhaus; earlier this concept was expressed by other terms: saturation capacity, atomicity, or basicity (Kekulé), division and units of affinity (A. W. Kolbé, E. Frankland, A. Butlerov, J. H. van't Hoff). Thus, the concept of affinity, which in fact referred to the ability of a substance to react, was in a certain period of the development of chemistry associated with the concept of valency. Some elements of the concept of valency can be found in Avogadro's hypothesis formulated in 1811 (but accepted only fifty years later). This held that one equivalent of oxygen is combined with two equivalents of hydrogen or chlorine, and one equivalent of sulphur. A significant step towards the concept of valency was Laurent's statement from 1846 that in each molecule the sum of odd-valency atoms, i.e. atoms of hydrogen, nitrogen, chlorine, boron, phosphorus, and arsenic must be expressed as an even number.

The credit for the formulation of the concept of valency is given by the historians of chemistry to Edward Frankland, notwithstanding the fact that valencies resulting from his formulae are in many cases twice as great as the real values. His application of equivalents instead of atomic masses (8 for oxygen and 6 for carbon) causes the assumed stoichiometric relations of atoms in molecules to be incompatible with the real ones. In 1852 Frankland reported the results of his experiments on organometallic compounds:

When the formulae of inorganic chemical compounds are considered, even a superficial observer is impressed with the general symmetry of their construction. The compounds of nitrogen, phosphorus, antimony and arsenic, especially, exhibit the tendency of these elements to form compounds containing 3 or 5 atoms of other elements; and it is in these proportions that their affinities are best satisfied: thus in the ternal group we have NO_3, NH_3, NI_3, NS_3, PO_3, PCl_3, SbO_3, SbH_3, $SbCl_3$, AsO_3, AsH_3, $AsCl_3$, etc; and in the five-atom group, NO_5, NH_4O, NH_4I, PO_5, PH_4I, etc.

And next:

Taking thus view of the so-called conjugate organic radicals, and regarding the oxygen, sulphur, or chlorine compounds of each metal as the true molecular types of the organometallic bodies derived from them by the substitution of an organic group of oxygen, sulphur, etc., we have following inorganic types and organometallic derivatives:

$As\begin{Bmatrix} S \\ S \end{Bmatrix}$ $As\begin{Bmatrix} C_2H_3 \\ C_2H_3 \end{Bmatrix}$ $As\begin{Bmatrix} O \\ O \\ O \\ O \\ O \end{Bmatrix}$ $As\begin{Bmatrix} C_2H_3 \\ C_2H_3 \\ O \\ O \\ O \end{Bmatrix}$

          cacodyl

$As\begin{Bmatrix} O \\ O \\ O \end{Bmatrix}$ $As\begin{Bmatrix} C_2H_3 \\ C_2H_3 \\ O \end{Bmatrix}$ cacodylic acid anhydrite

          oxide of cacodyl

$Sb\begin{Bmatrix} O \\ O \\ O \end{Bmatrix}$ $Sb\begin{Bmatrix} O \\ O \\ O \\ O \\ O \end{Bmatrix}$ $Sb\begin{Bmatrix} C_4H_5 \\ C_4H_5 \\ C_4H_5 \\ C_4H_5 \\ O \end{Bmatrix}$ $Sb\begin{Bmatrix} C_4H_5 \\ C_4H_5 \\ C_4H_5 \\ O \\ O \end{Bmatrix}$

                              stibenite oxide   stibenite dioxide

$Sb\begin{Bmatrix} C_4H_5 \\ C_4H_5 \\ C_4H_5 \end{Bmatrix}$

stibenite

ZnO $Zn(C_2H_3)$ $Zn\begin{Bmatrix} C_2H_3 \\ O \end{Bmatrix}$

               zinc methide   oxide of zinc methide

4.3. VALENCY

SnO   Sn(C$_4$H$_5$)   Sn$\begin{Bmatrix}C_4H_5\\O\end{Bmatrix}$

                    stannous ethide   oxide of stannous ethide

Hg$\begin{Bmatrix}I\\I\end{Bmatrix}$ Hg$\begin{Bmatrix}C_2H_3\\I\end{Bmatrix}$

               mercuric
               methiodide

[142].

It is evident from the formulae quoted above what valencies Frankland attributed to the particular atoms and groups of atoms. Besides atoms displaying a double valency (3 and 5), there also occur divalent and univalent atom of metals. In spite of many inaccuracies Frankland developed the type theory and showed more clearly than any preceding investigators that some atoms and groups of atoms can replace other atoms or groups of atoms in a one-to-one relation. In this way he introduced what we now call a 'unit of valency'.

Next, in 1854, Berthelot proved that trivalent groups C$_3$H$_5$ must occur in glycerol, and in 1855 Heinrich Buff wrote about bivalent radicals (groups). However, Kekulé made the greatest contribution to the development of the concept of valency in his two publications dating from 1857 and 1858. In the first publication he distinguished a group of univalent elements (monoatomic or monobasic, according to his terminology): H, Cl, Br, K; a group of bivalent elements: O, S; and trivalent elements: N, P, As. He also mentioned the tetravalency of carbon [143]. His second work has greater importance. Kekulé writes:

The radical of sulphuric acid SO$_2$ contains three atoms, each of which is diatomic, thus representing two affinity units. By joining together, one affinity unit of one atom combines with one of the other. Of the six affinity units, four are thus used to hold the three atoms themselves together: two remain over, and the group becomes diatomic; it unites, e.g., with two atoms of a monatomic element:

        Sulphuryl radical   Sulphuryl chloride

          S$\begin{Bmatrix}O\\O\end{Bmatrix}$ S$\begin{matrix}Cl\\Cl\end{matrix}\begin{Bmatrix}O\\O\end{Bmatrix}$

(...) In a similar way the manner in which the atoms are arranged can be shown for all radicals, including those which contain carbon. To do this, it is only necessary to form a picture of the nature of carbon. If only the simplest compounds of carbon are considered (marsh gas, methyl chloride, carbon tetrachloride, chloroform, carbonic acid, phosgene gas, carbon disulfide, prussic acid, etc.), it is striking that the amount of carbon which the chemist has known as the least possible, as the atom, always combines with four atoms of a monatomic, or two atoms of a diatomic element; that generally, the sum of the chemical unities of the elements which are bound to one atom of carbon is equal to 4. This leads to the view that carbon is tetratomic (or tetrabasic) (...). It simplest combinations with elements of the three other groups are

IV+4 I	IV+(II+2 I)	IV+2 II	IV+(III+I)
or, in examples,			
CH_4	$COCl_2$	CO_2	CNH
CCl_4		CS_2	
$CHCl_3$			
CH_3Cl			

For substances which contain more than one atom of carbon, it must be assumed that at least part of the atoms are held just by the affinity of carbon and that the carbon atoms themselves are joined together, so that naturally a part of the affinity of one for the other will bind an equally great part of the affinity of the other. The simplest, and therefore the most obvious, case of such linking together of two carbon atoms is when one affinity unit of one atom is bound to one of the other. Of the 2×4 affinity units of the two carbon atoms, two are thus used to hold both atoms themselves together; there still remain six which can be bound by the atoms of other elements (...) If we put more than two carbon atoms together in the same way, then for each further one added, the basicity of the carbon group will be raised by two units. The number of hydrogen atoms (chemical units) which is bound with n atoms of carbon joined together in this way will be expressed by $n(4-2)+2 = 2n+2$ [144, 145, p. 155].

Thus, Kekulé in 1858 derived a formula for a homologous series of aliphatic hydrocarbons. Further on in the work quoted he writes:

When comparisons are made between compounds which have an equal number of carbon atoms in the molecule and which can be changed into each other by simple transformations (e.g., alcohol, ethyl chloride, aldehyde, acetic acid, glycolic acid, oxalic acid, etc.) the view is reached that the carbon atoms are arranged in the same way and only the atoms held to the carbon skeleton are changed.

Drawing formulae of the compounds in the above-mentioned two works, Kekulé used brackets to show which atoms and groups are

4.4. RATIONAL FORMULAE

joined together. In the following year, Archibald Couper replaced the brackets commonly used in those days by dashes which joined particular atoms, thanks to which chemical formulae assumed a form that resembles the structural formulae which we use today.

The development of the concept of valency (regardless of the names given to it by various authors) on one hand contributed as we have said in the preceding chapter to the systematization of the elements and, on the other hand, it enabled the research on the chemical structure of complex organic compounds to be developed. The concept of valency was for dozens of years the main guideline in all chemical considerations. It was not until the last decade of the 19th century that the concept of valency proved to be insufficient, and no significant modification was made to it until as late as the 1930s.

4.4. Rational Formulae

As we have seen, there were two approaches to determine the structure of organic compounds: the historically older method of analysis and the subsequent method of synthesis. Both were based on the observation of reactions; hence formulae, particularly those obtained by the synthesis method, were closely connected with chemical reactions. However, fairly soon an opinion was expressed that these formulae can represent real atomic bonds in a molecule. Adherents of this opinion stated that they were seeking 'rational formulae'. It was Berzelius who already at that time held that formulae derived on the basis of the dualistic theory represent the actual atomic bonds in groups and bonds between groups of atoms. Consequently, it was held that in the aqueous solutions of acid oxides (now regarded as acids) there are always compound atoms (molecules) of water.

Also Gerhardt wrote quite explicitly in his textbook that his formulae represented molecular constitution, i.e. "a real arrangement of atoms" [140, vol. IV, p. 561], and according to the German chemist Adolf Kolbé "the chemical constitution is a real starting point for chemical forces of affinity" [146]. The terms which were then coined—chemical constitution or structure—were frequently treated as

4. THE STRUCTURE OF CHEMICAL COMPOUNDS

synonyms; even today, due to the development of organic chemistry, it is hard to prescribe their precise meanings.

During the 36th Congress of German Naturalists and Physicians in 1861, the Russian chemist Alexandr N. Butlerov defines the meaning of the term 'structure' in the following way:

> Assuming that in each chemical atom there is only a definite and limited amount of chemical force (affinity) which takes part in forming a body, I would refer the term chemical structure to a chemical compound (bond) or to the kind and manner of mutual bonding of atoms in a complex body [147].

Next Butlerov states that formulae which describe the chemical structure are:

> (...) not yet completely, though, at least to some degree, real rational formulae. Only one rational formula will be thus in this sense attributed to each body, and when the general laws of the dependence of chemical properties of bodies on their chemical structure are known, a similar formula will express all these properties.

In order to construct such formulae, Butlerov attributes to "an atom of each element a definite amount of force which causes chemical phenomena". All of this force or part of it is used to form a chemical bond. He is of the opinion that a unit of this force should be "a minimum of force contained in atoms of some elements, e.g. hydrogen". Then, "by comparing it with an amount of affinity specific to other atoms, we find that the ratio between them is always expressed by whole numbers" [147]. In this way Butlerov attempted to determine what we would call today a unit of affinity, although nobody at that time was capable of describing the mechanism of the phenomenon of valency. It should be noted, however, that Butlerov distinguished these 'units of affinity' from bond energy. We shall return to this problem in the next chapter.

It follows from what has already been said that as early as 1861 Butlerov considered chemical formulae to be a real representation of the manner of bonding of atoms in a molecule.

More than a dozen years later, in 1879, he defined the chemical structure more precisely as the arrangement of chemical bonds between atoms in a molecule [148]. In 1863 he presented significant

4.4. RATIONAL FORMULAE

conclusions derived from his approach in a work published in *Zeitschrift für Chemie und Pharmacie*. Analysing the phenomenon of isomerism, he criticizes the formula of acetic acid provided by Kekulé in his textbook:

$$\left.\begin{array}{c} CH_3 \\ CO \\ H \end{array}\right\} O$$

and proposes transposing the upper brace to the left-hand side of the formula in order to emphasize the fact that both carbon atoms are directly bonded with each other. However, neither Kekulé in 1859 nor Butlerov four years later recognized the structure of the carboxyl group —COOH.

Butlerov's approach to structural formulae, discussed above, enabled him to define the phenomenon of isomerism more precisely as the result of different successions in the mutual bonding of atoms. In his work of 1863 we find the following assertion:

(...) atoms of one element in a molecule can and ought to differ from one another, if their chemical dependence from other elements in the same molecule is different.

From numerous examples presented by the Russian chemist we shall quote only one pair; glycine ($H_2N \cdot CH_2COOH$) and methyl carbamate ($H_2N \cdot COOCH_3$). Butlerov proposed the following formulae for these two compounds [149]:

$$\begin{cases} CH_2, NH_2 \\ CO, HO \end{cases} \quad \text{and} \quad \begin{cases} CO, NH_2 \\ CH_2, HO \end{cases}$$

Kekulé represented another approach. Following Gerhard's unitary theory he wrote in his textbook in 1859 that:

(...) the rational formulae of chemists do not express the binding of atoms and they cannot do so; they rather represent interrelations, analogy, and reactions. They are not formulae of constitution but formulae of transposition. Rational formulae represent in a conventional way a certain number of reactions; they are a kind of summary expressions for reactions written in the form of equation [145, p. 92].

4. THE STRUCTURE OF CHEMICAL COMPOUNDS

In the next chapters Kekulé develops this view:

> The aim of rational formulae is to give an idea of the chemical nature of compounds, i.e. of their metamorphoses and relationships with other bodies. In order to achieve it, atoms are written in formulae in such an arrangement as to indicate which atoms can be easily exchanged for others, and which groups of atoms remain intact in certain reactions (...). For a majority of substances different rational formulae are possible, and even in many cases one rational formula may not indicate all metamorphoses [145, p. 153].

Further, Kekulé asserts:

> It must be of course borne in mind that rational formulae are only formulae of transpositions, and not formulae of constitution, that they do not express anything but metamorphoses and comparisons of different substances, that they do not express by any means the binding of atoms in the existing combinations [145, p. 157].

Kekulé believes that one of the causes is the impossibility of representing a real, spatial arrangement of atoms in a molecule on a sheet of paper.

Analysing the way in which the formulae of chemical compounds are written, it will be worthwhile mentioning another representation of atoms in the form of 'caterpillars'. Kekulé employed this technique in his textbooks published in the years 1859–1867 and so did Czyrniański in the years 1862–1884. The caterpillars were formed by connecting the rings which represented the valency of atoms inside a common envelope. Thus a univalent hydrogen atom was represented by a ring, but an oxygen atom by a caterpillar with two lobes, and a caterpillar representing a quadrivalent carbon atom had four lobes. Kekulé employed caterpillar formulae mainly for organic compounds, but sometimes also for inorganic compounds. For example, he proposed the following formulae for sulphur dioxide and methane:

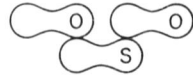 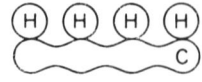

For convenience, Kekulé later replaced caterpillars by rods of the same length, whereas Czyrniański remained faithful to caterpillars until the end of his life. This was due to the more profound

4.5. UNSATURATED AND AROMATIC COMPOUNDS

significance which the Polish scientist attributed to caterpillars. Every atom—according to Czyrniański—has a characteristic amount of eddy motion whose unit is the angular momentum of a hydrogen atom, and the valency of particular atoms depends on the free units of angular momentum which are contained in them. In Czyrniański's theory we may find certain analogies to the currently assumed role of electron spins in the formation of a chemical bond; however, Czyrniański attributed angular momenta to atoms (and their parts) without any sufficient justification. Besides, his theory seems to have been premature, and the author, due to insufficient consideration of his theory, made errors in his assumptions. These errors cannot be justified by any inadequate knowledge of the laws of the microcosm [150].

4.5. The Structure of Unsaturated and Aromatic Compounds

A further step in the development of the conception of the structure of compounds was the introduction of the concept of multiple bonds. In 1863, when analysing the structure of azobenzene, Emil Erlenmeyer assumed that both nitrogen atoms are interconnected in this compound by a double bond. Two years later, he extended the notion of multiple bonds to bonds in some carbon compounds. In 1865, at the meeting of the Paris Chemical Society, Kekulé also spoke about such a possibility in his paper concerning the structure of aromatic compounds:

If a large number of carbon atoms are combined, it may first of all take such a course that a single unit of affinity of one atom is combined with one unit of affinity of a neighbouring atom (...). It may be further assumed that several atoms of carbon are arranged in a series in such a way that they constantly combine through every two units of affinity. Next it may be assumed that a bond occurs alternately through one and through two units of affinity. The first and last of these assumptions can be represented by the following periods:
1/1; 1/1; 1/1; 1/1, etc.
1/1; 2/2; 1/1; 2/2, etc.

The first kind of arrangement of carbon atoms accounts for homology, and besides the structure of fatty (aliphatic—R.M.) bodies. Acceptance of the arrangement according to the second law of symmetry leads to the explanation

184   4. THE STRUCTURE OF CHEMICAL COMPOUNDS

of the structure of aromatic substances or at least to the core which is common for these substances.

If, namely, we assume that six carbon atoms are arranged according to this law of symmetry, then we shall obtain a group which when considered as an open chain, as is illustrated by the first of the graphic formulae presented below, has still eight unsaturated units of affinity. On the other hand, if we assume further that either carbon atom which completes the chain, is interconnected by one unit of valency, as shown by the arrows of the second graphic formulae, then we have a closed chain (a symmetric ring) which contains six free units of affinity

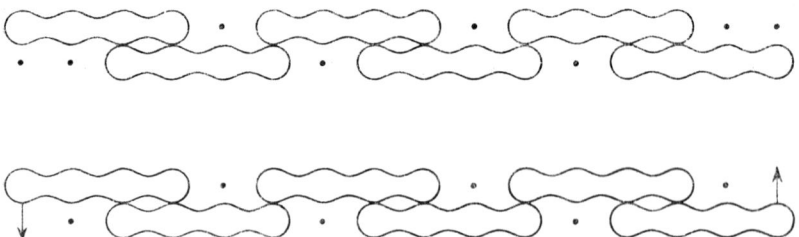

The opinion concerning the structure of this closed chain consisting of six carbon atoms will be expressed even more explicitly by the following graphic formula in which carbon atoms are represented by circles and the four units of affinity by four lines leading from them

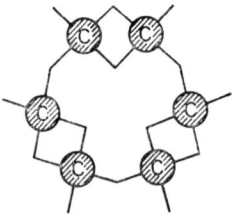

All compounds which are usually referred to as aromatic substances (...) are derived from this closed chain [151].

At the ceremony of the 25th anniversary of the theory of the benzene ring, Kekulé spoke about his first vision of ring arrangement which he had while working on his textbook in Ghent, Belgium. While resting in a drowse he saw spinning atoms being arranged in serpentine chains. At some moment he saw a serpent catch its own tail and begin to whirl violently. According to the opinion of the Swiss historian of

4.5. UNSATURATED AND AROMATIC COMPOUNDS

chemistry H. E. Fierz-David, that vision resulted from an association with the alchemical symbol of ouroboros (the serpent that devours its own tail), which we have discussed in Chapter 1.

The representation of various organic molecules in the form of polygons (most frequently hexagons) can be found in Laurent's textbook [90] published as far back as 1854. Some of the sides of those figures represent individual atoms, although the number of sides is not equal to the number of atoms in the molecules.

The ring-like arrangement of carbon atoms in aromatic compounds was soon accepted by the chemists. The main argument supporting this hypothesis was the fact that in the case of an open chain not all hydrogen atoms would be equivalent, therefore there should be a few monosubstituted isomeric derivatives of benzene for each compound. Only one isomer of such compounds was known. This proved the equivalence of all the six free units of affinity; it was possible only for the ring-like arrangement of carbon atoms.

On the other hand, single and double bonds, arranged alternately in the ring, might suggest the existence of two isomers of the derivative in which two neighbouring atoms of hydrogen are replaced by two atoms of another element; the two atoms would substitute for hydrogen atoms attached to the carbon atoms fixed by a single bond in one isomer, and by a double bond in the other isomer.

However, nobody seriously considered the possibility of the existence of such isomers. In his textbook of 1867, Kekulé proposes representing molecule of benzene by a regular hexagon in which all the bonds might be marked by single dashes [152, p. 22]. At the same time, referring to P. Havrez's textbook *Principes de la chimie unitaire* published in 1866, he suggests a spatial arrangement of atoms, which conforms with the opinions of the same investigator concerning the spatial structure of molecules presented earlier in this book.

In a textbook published in 1867, Kekulé predicted the existence of three isomers of benzene derivatives substituted in positions 1,2, 1,3 and 1,4. The appropriateness of this assertion was confirmed in the years 1869–1874 by Kekulé's pupil Wilhelm Körner who called them ortho, meta, and para isomers. Kekulé also predicted the existence of three- and four-substituted isomeric derivatives of benzene. Hence it follows that Kekulé always recognized in practice the equivalence of all bonds between carbon atoms in a ring despite the fact that the starting point of his considerations was a chain of alternately arranged single and double bonds. Such a ring would correspond to cyclohexatriene, but benzene behaves like cyclohexatriene only in reactions in which atoms are attached to the two neighbouring atoms of carbon; the whole structure of the ring is then changed.

However, in trying to explain this equiponderance of the bonds in a ring, the chemists mentioned two cyclohexatriene rings which supplemented each other and differed by the place of the double and single bonds. They also proposed numerous other structures for the ring. A definite solution was provided in 1899 by the notion of Johann Thiele, according to which there occurs a certain equalization of the product of bonding in all systems having alternately-coupled double and single bonds. An understanding of this problem was possible only after the role of electrons in the formation of chemical bonds had been explained.

Analysing the approaches of Kekulé and Butlerov to the rational formulae, we can notice a significant difference; Kekulé wanted first of all to facilitate an understanding of the course of chemical reactions; he intended to describe the structure of molecules in a state just before the reaction, when they remained, so to speak, under the influence of the reagent with which they were to react. Butlerov, on the other hand, attempted to present an idealized structure of an isolated molecule; this structure—as we know today—undergoes modifications under the influence of the environment.

4.6. Stereochemistry

As we have seen, the scientists who proposed rational formulae of molecules frequently pointed out the role of the spatial arrangement of atoms. This idea goes back to optical investigations carried out from the first decade of the 19th century. In 1808, the French physicist Etienne Louis Malus was the first to observe that light can be linearly polarized by reflection, i.e. one can obtain a light beam whose

4.6. STEREOCHEMISTRY

electrical vector—according to modern views—always oscillates in one and the same plane. In 1815, another French physicist, Jean Baptiste Biot, stated that some substances, such as sugar, camphor, and tartaric acid, rotate the polarization of a linearly polarized light beam when it passes through them. Biot noticed that some of these substances lose this property after melting, and others retain it even after dissolution in water. He also noticed that sometimes these substances, today referred to as 'optically active' or 'chiral', when subjected to chemical reactions, continue to rotate the plane of polarization of light whereas, as a result of other reactions, they lose this property.

In 1821, Sir John Herschel proved that hemihedral varieties of quartz crystals, which are a mirror reflection with respect to each other, rotate the plane of polarization by an equal angle, but in opposite directions. We have already mentioned that in 1830 Berzelius showed that the composition of optically active tartaric and that of optically inactive racemic acid were identical. In 1848, the crystals of sodium-ammonium tartrate were investigated under a microscope by Louis Pasteur. He found that in fact he was examining a mixture of two kinds of crystals, one kind being a mirror image of the other. After a separation of crystals of a dual kind, he obtained from the one kind a tartaric acid which rotates the polarization plane of polarized light to the left, and from the other a compound that produced a rotation to the right [153]. The latter kind of acid had already been known for a long time. Although it was separated in the pure state by Scheele as late as 1769, its potassium salt, known as *sal tartaricum*, was obtained in antiquity in the process of the distillation of wine from mash. Laevorotary tartaric acid was first obtained by Pasteur as a result of the separation of crystals just described. He commented on that discovery as late as 1860:

We know, on the one hand, that the molecular structures of the two tartaric acids are asymmetric, and on the other, that they are rigorously the same, with the sole difference of showing asymmetry in opposite senses. Are the atoms of the right acid grouped on the spirals of a dextrogyrate helix, or placed at the summits of an irregular tetrahedron (...), or disposed according to some particular asymmetric grouping or other? We cannot answer these questions. But it cannot be a subject of doubt that the arrangement of atoms corresponds to an asymmetric system which does not superimpose on its mirror image [154].

4. THE STRUCTURE OF CHEMICAL COMPOUNDS

Thus in 1860 Pasteur propounded the possibility of the tetrahedral structure of the molecules of tartaric acid (asymmetry referred to the whole molecule). Similar suggestions were proposed by Kekulé in 1867 and the Italian chemist Emanuelo Paterno in 1869. However, these suggestions were made *ad hoc* in order to account for the observed facts. The actual proof of the tetrahedral structure of the carbon atom was carried out by Jacobus Hendricus van't Hoff in 1874 and independently of him, two months later, by Joseph Achille Le Bel.

Van't Hoff presented his theses first in the Dutch journal *Archives Néerlandaises des sciences exactes et naturelles*, and because this journal was little known, in the following year he published in French a pamphlet entitled *La chimie dans l'espace* (Chemistry in Space). At the outset van't Hoff defines the new concepts:

It was organic chemistry—carbon chemistry—that gave birth to this beautiful theory of atomicity which enables molecules to be presented as a grouping of atoms interconnected with one another in accordance with definite laws, which forms a complete and stable system.

Next, the author defines the objectives of a structural theory which should solve the problem of atomic positions with respect to each other and their motions. And since the motion of atoms alters their position, it is necessary to analyse positions "in one phase of their motion".

Van't Hoff was not satisfied with the structural formulae which had been used until his time; he considered that they were not sufficiently precise. By means of coloured models he accurately compared the possibility of different distributions of groups substituting hydrogen atoms in a methane molecule. It follows from his considerations that, assuming a flat formula for a methane molecule with the four valencies of a carbon atom pointing towards the vertices of a square, the following numbers of isomers of methane derivatives are possible:

one isomer for CH_3R and CHR'_3
two isomers for $CH_2R'_2$ and CH_2RR'''
three isomers for $CHR'R''R'''$ and $CR'R''R'''R^{IV}$

However, it followed from the experiments that the numbers of isomers of methane derivatives were smaller. Van't Hoff obtained

4.6. STEREOCHEMISTRY

compatibility with the experiments when he assumed that the affinities of a carbon atom are directed to the vertices of a tetrahedron whose centre is occupied by this atom. The predicted and detected number of isomers is then:

one isomer for CH_3R', $CH_2R'_2$, $CH_2R'R''$, CHR_3, CHR'_2R''
two isomers for $CHR'R''R'''$ or $CR'R''R'''R^{IV}$.

In other words: "In the case when the four affinities of a carbon atom are saturated with four groups differing from one another, we can obtain two and only two different tetrahedra, one of which being a mirror image of the other, and one can never superimpose them in mind, i.e. that we have to do with two isomeric formulae in space" van't Hoff concludes and illustrates this statement with diagrams of two such 'tetrahedra'

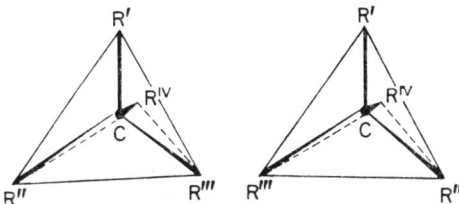

A carbon atom combined with four different substituents was called by van't Hoff 'an atom of asymmetric carbon'. By the presence of such asymmetric atoms he explains the phenomenon of rotation of the polarization plane of light. Furthermore, the presence of such an atom would be a cause of appearance of two isomers, which are their mutual mirror images.

Such considerations led van't Hoff to two conclusions:

i. A compound of saturated carbon which rotates the polarization plane of light in solution contains an atom of asymmetric carbon.

ii. The derivatives of optically active compounds lose their property of rotation when the carbon atom asymmetry disappears; in an opposite case they, as a rule, do not lose this property [155].

The aliphatic carbon chain is formed, in accordance with van't Hoff's model, when the tetrahedra of carbon atoms contact one

190   4. THE STRUCTURE OF CHEMICAL COMPOUNDS

another by the vertices. This model also allows prediction of the numbers of isomers in compounds containing different numbers of asymmetric carbon atoms. Van't Hoff writes further in his work:

I. One atom of asymmetric carbon in any carbon compound with a single carbon bond permits prediction of (existence) two isomers.

II. Every new atom of asymmetric carbon increases this number; it amounts to four when we have to do with two asymmetric carbons.

III. A compound containing n atoms of asymmetric carbon permits to predict existence of $2n$ optical isomers, whereas the former theory predicted only one (isomer).

The hypothesis of a tetrahedral carbon atom also enabled the Dutch chemist to present the double bond in a new light:

The double bond is represented by two tetrahedra with one common edge in which AB represents a combination of two carbon atoms, and R', R'', R''', R^{IV} represent univalent groups which saturate the remaining free affinities of carbon atoms.

By means of tetrahedra van't Hoff also constructed a model of the benzene molecule, both as a ring structure—in accordance with Kekulé's conception—and as a prism structure with a triangular base—in accordance with Ladenburg's suggestions. The triple bond was represented by van't Hoff in the form of two tetrahedra sharing a common face.

The concepts presented above were applied to ring compounds by Adolf Baeyer in 1885. He postulated a flat structure for the rings made up of carbon atoms. In a review of the contemporary state of the art he asserted that a carbon atom has four equivalent valences uniformly distributed in space, and that carbon atoms can combine with one another by the formation of a single, double, or triple bond, thereby forming open or closed chains. On this basis he stated further:

I should like to add to these generally acknowledged truths what follows: Four valences of a carbon atom act in directions which bind the centre of a sphere with the vertices of a tetrahedron and contain an angle 109°28'. This direction of attraction can be deflected, which causes stress increasing with the magnitude of deflection [156].

The above supplement includes so-called 'Baeyer's theory of stresses' which states that the arrangement of carbon atoms is the less stable the more different are the angles between the bonds of successive carbon atoms from the angle 109°28′, which is characteristic of the tetrahedron. Therefore, of all the flat ring arrangements five- and six-member rings are most stable according to Bayer.

Although van't Hoff and Le Bel were not the first investigators to propose the concept of a tetrahedral carbon atom, they were the ones who proved it eventually. In spite of the reluctance of some chemists, Kolbe among them, this concept soon became widespread. It was Johannes Wislicenus who contributed greatly to its dissemination. On the basis of this theory he explained *cis-trans* isomerism by the examples of pairs of fumaric and maleic acids as well as crotonic and isocrotonic acids, and later he consistently made use of this theory in his textbook of organic chemistry, published in 1887.

Three years later, Artur Rudolf Hantzsch and Alfred Werner explained the isomerism of oximes by means of the spatial structure of molecules, which could not have been understood on the basis of the flat formulae, and also the isomerism of azo compounds, and they predicted the isomerism of tertiary amines. The spatial isomerism of the two latter types of compounds was based on a representation of nitrogen atoms as pyramids with a triangular base. Next Frederic Stanley Kipping, Smiles, and Jackson Pope, independently of each other, turned their attention to the spatial isomerism of the compounds of sulphur, selenium, bismuth, silicon, and phosphorus. Besides, Werner investigated the spatial isomerism of the compounds of cobalt, chromium, platinum, and some other metals.

4.7. Conformational Analysis and Rotational Isomerism

When investigating the structure of organic compounds Wislicenus wondered whether the principle of free rotation around a single bond between the atoms of a carbon chain, which results from the combination of the tetrahedra of carbon atoms, always holds true. This problem was solved by his pupil Carl A. Bischoff, who proposed a hypothesis of isomers differing by a spatial distribution of groups linked by a single bond. Bischoff called such isomers 'dynamic'. However, the conceptions of such a type of isomerism, today referred to as 'rotational isomerism', were only developed as late as post-1930. They were based on the fact that particular atoms in chemical compounds sometimes are not wholly neutral, but carry a certain fractional electrical charge. This is due to the role of electrons in the formation of a chemical bond. Atoms with a fractional charge of the same sign repel one another; therefore energetically it would be most advantageous to arrange them at the greatest possible distance from each another. The first experimental clue about the existence of rotational isomers (conformers) was provided by the research on the temperature-dependence of the capacity to rotate plane of polarization of light of halogen derivatives of ethane of the type ZYXC-CXYZ, carried out in 1928 by R. Lucas, as well as by the changes with temperature of the dipole moment of 1,2-dichloroethane observed in 1932 by K. Zahn. Convincing evidence of the existence of two isomers of this compound with different spatial symmetry was given by K. W. F. Kohlrausch's investigations on the spectra of the Raman effect. Knowledge of both the number of atoms making up a molecule and its symmetry allows calculation of the number of Raman bands which should occur in the spectrum. However, the number of bands of 1,2-dichloroethane observed by Kohlrausch was almost twice as great. The investigator explained this observation by assuming that this compound occurs in two forms of different symmetry. In 1939, K. S. Pitzer calculated by thermodynamic methods that in a molecule of ethane the most probable arrangement of the two $-CH_3$ groups with respect to each other is that in which—when we look at it along the bond C—C—the hydrogen atoms of one group are visible between the hydrogen atoms of the other group. The difference in energy between this stable system and an unstable system in which hydrogen atoms of both groups are exactly opposite to one another equals 656 J/mol [157].

In the case of 1,2-dichloroethane there exist two stable conformers: the one in which chlorine atoms are detached from one another as far as possible, was formerly referred to as a staggered isomer by analogy

4.7. CONFORMATIONAL ANALYSIS

with the isomerism around the double bond; the other, in which the planes of both C—C—Cl systems form an angle of 60°, was referred to as a skew isomer. And, finally, the unstable conformer, in which planes of C—C—Cl systems in Newman's projection are superimposed, was referred to as an eclipsed isomer. At present the staggered and eclipsed isomers are called 'sinperiplanar' and 'antiperiplanar' conformers, respectively, and the eclipsed isomers are called 'sinclinal' conformers.

Before the concepts of rotational isomerism were finally formed, some chemists had developed the conformation analysis of the cyclohexane derivatives. In 1890, H. Sachse, under the influence of Bischoff's considerations, came to the conclusion that cyclohexane should occur in two isomeric forms. One, with a double symmetry axis and two intersecting symmetry planes, has a form resembling that of a boat; the other, with a triple symmetry axis and a symmetry centre, resembles a chair. These isomers could not be separated because—as C. Mohr pointed out in 1918—their energies are very close to each other, the energy barrier between them being very small.

In 1938, O. Hassel proved that cyclohexane must exist in two isomeric forms as the full number of its derivatives could not be explained only on the basis of substitution of hydrogen atoms of the particular CH_2 groups. For example, 1,2,3,4,5,6-hexahydroxycyclohexane (inositol) occurs in several forms of which one has the property of rotating the plane of polarized light although there is no asymmetric carbon atom, and each of the carbon atoms present is connected with the two —CH_2— groups. In the years 1948–1950 Derek H. Barton developed Sachse's ideas and distinguished a new branch in chemistry called 'conformation analysis'. He showed that the particular conformational isomers—conformers—which result from inhibition of the free rotation of atoms around the single C—C bond, have different electronic, oscillation and rotation spectra, give different diffraction patterns of the dispersed electron beams and X-rays, different spectra of nuclear resonance and different optical rotatory dispersions. The rates of chemical reactions of particular conformers are also different.

4.8. The Theory of Ions

Electrochemical reasearch, which developed from the early 19th century, contributed to the explanation of some aspects of the mechanisms of chemical reactions (cf. Chapter 5) on one hand, and the structure of some chemical compounds and their behaviour in aqueous solution on the other. The laws of electrolysis formulated by Faraday in fact revealed, as we have said, the shortcomings of the dualistic electrochemical theory of Berzelius, but at the same time they also showed that the molecules of acids, bases, and salts can be decomposed into electrically charged constituents—ions. In the process of electrolysis ions were neutralized on the electrodes, but it was not clear at that time where and when they are formed. In 1857, Rudolf Clausius proved in an extensive article that this decomposition cannot occur only near the electrodes because current conduction through electrolytic solution would not be in conformity with the known and experimentally verified laws of electric current flow. Thus, in accordance with the kinetic theory of matter, molecules must be—in his opinion—in a state of dynamic dissociation into electrical parts in the whole volume of the solution, whereby in turn these parts become combined with one another [158]. Clausius does not quantitatively analyse the equilibrium state between the dissociated and complete molecules because the notion of chemical equilibrium was formulated only ten years later (cf. Chapter 5).

In 1881, H. Helmholtz, while considering the process of electrolysis, came to a conclusion that "no other force can resist the motion of ions through a liquid but a mutual attraction of their electrical charges" [133]. This conclusion is now regarded as an introduction of the concept of the ionic bond.

In the next few years the properties of dilute aqueous solutions of electrolytes were investigated by Wilhelm Ostwald of Germany and Svante Arrhenius of Sweden. Ostwald stated that the molar conductance of these solutions increases with the dilution of electrolyte and reaches a certain limiting value for infinitely dilute solutions. He also confirmed a hypothesis, proposed earlier by F. W. Kohlrausch, that the motions of ions with opposite signs formed by the de-

4.8. THE THEORY OF IONS

composition of the electrolyte molecules are independent of one another [159]. On the basis of his investigations Arrhenius came to the same conclusion. He called the undecomposed electrolyte molecules 'inactive' and the decomposed ones 'active'. He also introduced the concept of the degree of dissociation, equal to the ratio of the number of active molecules to the sum of both kinds of molecules [160].

From 1885 Jacobus Hendricus van't Hoff was engaged in the study of the osmotic pressure of solutions. He pointed out that this pressure is equal numerically to the gas pressure when the number of gas molecules of a substance in equal volume and dissolved at equal temperature is the same. Similarly, the temperature relationship of osmotic pressure proved to be identical with that of gas pressure [161]. Based on a far-reaching analogy of these mathematical relationships, in 1887 van't Hoff explained erroneously (as we know today) the mechanism of the origin of osmotic pressure by the impact of the molecules of the dissolved substance on the impermeable barrier which separated the solution from the pure solvent. (This barrier is, however, permeable for the solvent in both ways.) It is thus the same mechanism which explains the pressure exerted by a gas on the vessel walls [162]. In 1901, he realized, however, that such an explanation was not wholly adequate because when he received the Nobel Prize in Stockholm as the first chemist in history, he asserted that "irrespective of any hypothetical conceptions about the cause of this pressure (...) the laws concerning it have the same form as the gas laws" [163]. This analogy between osmotic and gas pressure played an important role in the development of research on electric conductance of electrolyte solutions, and later it became a basis of the original theory of electrochemical cells formulated by W. Nernst [164].

This analogy allowed an explanation of deviations from the laws of osmotic pressure for electrolyte solutions which was observed by van't Hoff. This investigator showed that in the case of these solutions the osmotic pressure calculated for a given number of electrolyte molecules must be multiplied by a certain coefficient which he denoted by the letter i [162]. In the same year 1887, Arrhenius pointed out that deviations in the osmotic pressure of electrolyte solutions, observed by van't Hoff, are analogous to those of the pressure of halogens heated to a high temperature. These latter deviations can be accounted for by a decomposition of diatomic molecules into single atoms. Thus, according to Arrhenius, deviations in the osmotic pressure of electrolyte solutions are due to the decomposition of their molecules into ions. To that conclusion Arrhenius came when he derived a mathematical dependence of van't Hoff's coefficient i on the degree of dissociation [165]. He proved an agreement between the values of van't Hoff's coefficient determined for particular solutions by different methods.

The measurements of osmotic pressure were made without applying any electrical field. Arrhenius' publication therefore showed that ions are formed as a result of the dissolution of electrolyte in water. This confirmed a conclusion formulated by Ostwald as early as 1885 [166], when he studied the effect of dilution on the electric conductance of electrolyte solutions, that (electrolytic) dissociation is caused only by water and not by electric current [165].

The significance of the theory of electrolytic dissociation cannot be presented better than was done by W. Ostwald and W. Nernst in 1889: "As a concordant result of a great number of different research methods it becomes evident that it is not—as has been presumed so far—a negligibly small part of the electrolyte that undergoes dissociation but, on the contrary, it is only a small part that remains undissociated in a solution whereas its major part is usually dissociated into ions, particularly in the majority of neutral salts as well as in strong acids and bases" [167].

4.9. Extension of the Concept of Valency: Complex Compounds

After the crystallization of the concept of valency a discussion followed whether a given element always has the same valency, or whether it can vary. A. Kekulé was an adherent of only one valency for an element, although the considerations of Frankland that we have quoted above seem to speak for two valences of certain elements. It is worth mentioning that the Polish chemist E. Czyrniański, in the years 1862–1872, developed a concept that he was unable to justify: that atoms

4.9. EXTENSION OF THE CONCEPT OF VALENCY

consist of two parts with what we would call today quantified angular momenta which cancel in pairs, but the number of these cancelled pairs depends on the way in which the two parts of a given atom are united. Thus, the valency of such an atom could have different values, but it always had to be odd or even [168].

Next, in 1880, Otis Coe Johnson defined an 'oxidizing agent' as one that "increases the number of bonds of a substance (...). The number of bonds obtained by one (substance) is lost by the other" which is then reduced [169]. As a matter of fact Johnson does not write about valency, but changes in the number of bonds of a given agent cannot be understood otherwise.

The conception of ionic bonds developed by Helmholtz together with the theory of electrolytic dissociation identified the concept of valency of atoms or groups of atoms with the number of electric charges which the created ions had; this number was later named electrovalency.

The concept of equivalent began gradually to lose its significance. W. Ostwald, in a textbook published at the turn of the 19th century, introduced the concept of mole, and in A. Smith's textbook 'General Chemistry for Colleges' published in 1905 we find a definition of molar (atomic) mass as a product of equivalent A and valency W.

Another group of investigations which extended the concept of valency included studies on the structure of alums and compounds containing molecules of water, ammonia, cyanogen, and carbon oxide—today called hydrates, ammoniates, cyanide and carbonyl complexes. In 1899, Ostwald called them 'complex compounds', and recognized that their structure is analogous to that of organic compounds, as the latter also contain some definite, non-ionic atomic groups. We have already mentioned that Werner employed a stereochemical approach to the structure of inorganic compounds. From this point of view he investigated ammoniates and found that these compounds dissociate into simple chlorine ions and very complex cations which cannot be dissociated into further ions.

Analysing the structure of complex cations, which he called 'complex ions', in 1891 Werner provided his definition of the valency of an element and, precisely, its valency number:

4. THE STRUCTURE OF CHEMICAL COMPOUNDS

Affinity is an attractive force which acts uniformly from the centre of an atom towards all parts of its spherical surface. Valency means numerical ratios found empirically in which atoms combine with one another [170].

Werner introduced the term 'valency points' to denote the intersection points of the spherical surface of the atom with lines connecting the centres of gravity of atoms.

Two years later the Swiss author extended the definition of valency:

The valency number denotes the maximal number of atoms which can be directly connected with a given atom without the intervention of other elemental atoms.

The valency number concerned only the empirically determined values for common inorganic compounds. However, an atom of metal in a complex ion attracts a greater number of atoms and groups than might result from its valency number. Therefore, Werner introduced the concept of coordination number which means "the maximal number of monovalent atoms and groups being capable of remaining in a direct bond with a given atom". Thus the coordination number is a sum of the valency number, which he called 'primary valencies', and a certain number of other valencies, referred to as 'secondary valencies'.

However, the investigator does not make a sharp distinction between these two kinds of valency. He writes further in his work:

All attempts at marking a sharp boundary between the forces of primary and secondary valencies are futile. In reality the forces of primary and secondary valency are very similar or perhaps even identical. It becomes more and more evident with time that the distinction of the valency forces into primary and secondary is merely a temporary solution which is necessary at the present transient period of the science of valency [171].

Werner attributed to ammoniates the octahedral structure at the centre of which is a metallic atom or ion with the coordination number six. He accounted for the appearance of optical isomers of these compounds by a different arrangement of chlorine atoms which make up a complex ion, in a similar way as the existence of optical isomers of methane derivatives substituted by four different groups was explained. In his history of chemistry, Giua sums up Werner's contribution in the following way [9, p. 319]:

1. He drew a distinction between primary and secondary valencies.
2. He introduced the concept of coordination number and defined its maximum value.
3. This maximum value accounts for the maximum number of groups which can be arranged in a sphere surrounding the central atom.
4. Werner's theory describes the spatial arrangement of groups directly bonded with the central atom.
5. The theory describes the reaction capacity of groups and atoms bonded directly.

The concept of secondary valency also referred, of course, to all hydrates and carbonyls.

4.10. Electrons and the Chemical Bond

A more precise explanation of the mechanism of the formation of chemical bonds described both by primary and secondary valencies became possible in the 20th century thanks to the development of new physical conceptions. Of particular significance was J. J. Thomson's conclusion, dating from 1898, according to which electrons were inside the atom. In the next year Richard Abegg and G. Bodländer introduced the concept of electron affinity. They stated: "Since for the existence of inorganic compounds the affinity of atoms or groups for an electrical charge has a more decisive importance than the affinity of atoms for one another it will be useful to take the affinity of atoms and radicals as a basis for the taxonomy of inorganic bonds" [172]. In another work (1904), Abegg introduced the concept of electrovalency and showed that "every element is endowed with a positive and negative maximal valency, where the sum of positive and negative units of valency always equals eight" [173]. This formulation has been referred to as Abegg's theory of valency. Abegg also proved that the arrangement of eight electrons in the noble gases must be particularly stable. Abegg's theory proved more precisely Mendeleev's observation made 9 years earlier that the sum of valencies in respect of oxygen and hydrogen equals eight.

4. THE STRUCTURE OF CHEMICAL COMPOUNDS

In 1907, J. J. Thomson proposed a hypothesis that electrons take part in the chemical bond, and in 1910 K. G. Falk and J. M. Nelson in a work entitled 'The Electronic Conception of Valence' [174] suggested that during the formation of a bond electrons flow from one atom to another. Also in 1910 S. W. Dajn spoke (in an unpublished lecture) about the displacement of electrons. According to the account of G. Dajn published after four years [175], S. W. Dajn emphasized that "in every reaction of oxidation and reduction electrons move away from one atom and pass to another, their valency changing, one them being oxidized and the other reduced". The above mentioned publication of G. Dajn from 1914 passed unnoticed although it was extensively summarized—admittedly during the First World War—in *Chemisches Zentralblatt* (1915 I 867).

In 1914, J. J. Thomson took Abegg's theory into account of valency and proved that:

(...) the electrical field of the atom is not fully compensated by the field of negative charges since these charges are in different parts of the atom. This uncompensated field interacts with the atoms of the neighbouring molecules and becomes the source of such phenomena as pressure caused by a real gas—lower than it might result from the laws of a perfect gas, surface tension, vaporization heat, and cohesion.

Today these phenomena are attributed to intermolecular interactions

... whereas chemical affinity and chemical phenomena in general are the effects of forces having the same origin but acting between atoms of the same molecule [176].

Analysing the distribution of electrons in atoms and treating the electric system inside a molecule as a whole, Thomson proposed in 1914 the division of compounds into dipolar and non-dipolar ones. This conception was given a mathematical form in the years 1920–1921 by Peter Debye and Willem Keesom and led to the development of so-called van der Waals intermolecular interactions [177].

In his work of 1914, Thomson also provides a definition of a molecule:

In order that a chemical compound may exist in a stable form it must satisfy certain conditions: one of these is that its molecules must not exert on other

4.10. ELECTRONS AND THE CHEMICAL BOND

molecules in their neighbourhood attractions large enough to cause the molecules to unite and thus form another system; another condition is that the attractions between atoms in the molecule must be great enough to hold them together in spite of blows they receive when one molecule collides with another molecule.

Next, referring to Abegg's theory of valence, Thomson asserts that a certain number of moving electrons, less than eight, forms a ring around the stable core, the number of these moving corpuscles in the atom of an element being equal to the number of the group which this element occupies in Mendeleev's System. These electrons have no fixed place in an atom unless "a beam of force lines (emerging from such an atom—R.M.) hitches with its end something else than the same atom, i.e. it must terminate in another atom".

Analogous ideas to that of Thomson, but more precise, can be found in Gilbert N. Lewis' publication of 1916. He summarizes his views in the following points:

1. In every atom is an essential kernel which remains unaltered in all ordinary chemical changes and which possesses an excess of positive charges corresponding in number to the ordinal number of the group in the periodic table to which the element belongs.

2. The atom is composed of the kernel and an outer atom or shell, which, in the case of the neutral atom, contains negative electrons in number to the excess of positive charges of the kernel, but the number of electrons in the shell may vary during chemical change between 0 and 8.

3. The atom tends to hold an even number of electrons in the shell, and especially to hold eight electrons which are normally arranged symmetrically at the eight corners of a cube.

4. Two atomic shells are mutually interpenetrable [178].

Next Lewis states that the electrons within this shell can pass from one place to the other; he considered this concept to be more correct that the concept of electronic orbits proposed by Niels Bohr.

Thus Lewis conceived of atoms as cubes containing one to eight electrons at their corners. If a chemical bond is made the cubic atoms touch each other with edges in the case of a single bond, and with faces—in the case of a double bond, i.e. when two or four electrons coming from two cubes become common to two bonded cubic atoms. In the case of small atoms, e.g. carbon atoms, these vertical points can approach each other in pairs through magnetic attraction perhaps and in this way four vertices are made of eight vertices, i.e. a tetrahedron is formed. In each vertex of this new tetrahedron, identical with that postulated by van't Hoff, there may occur two electrons if a bond is formed.

Lewis' theory was developed three years later by Irving Langmuir [179] who pointed out that the interacting atoms tend to make up their shells to the stable configurations predicted by Lewis. Thus attachment of the missing electrons for completion of octet causes formation of a negative ion, whereas a return of the electrons revolving outside the stable configuration provokes the formation of a positive ion. Ions with opposite signs interact by means of electrostatic forces. Completion of octets can also occur, in accordance with Lewis' theory, by sharing the electrons coming from each of the bonded atoms; one, two, or three electron pairs can be involved. The considerations of Langmuir contributed to the refinement of the difference between the conception of the ionic bond introduced by Helmoltz and the conception of the covalent bond introduced by William Ramsey. The explanation of the formation of the ion bond by donation and abstraction of electrons was in a way a renewal of Berzelius' dualistic theory, although it did not yet include any quantitative measure of the capacity to donate or abstract electrons.

In 1923 G. N. Lewis explained the acidic and basic properties of compounds by their capacity to transfer electrons. He stated:

A basic substance is one which has a lone pair of electrons which may be used to complete the stable group of another atom; and an acid substance is one which employs a lone pair from another molecule in completing the stable group of one of its own atoms [180].

As can be seen, the mechanism of acid—base reaction presented by Lewis is in reality identical with the mechanism of oxidation—reactions proposed by S. W. Dajn.

In the same year, 1923, a symposium on the electronic theory of valence was held in London by the Faraday Society. G. N. Lewis played an active role in that symposium. He pointed out that there is a contradiction between the physical model of the atom proposed by

4.10. ELECTRONS AND THE CHEMICAL BOND

Rutherford–Bohr–Sommerfeld, according to which electrons have no fixed position but move around circular or elliptical orbits, and the chemical model of the octet, according to which the positions of electrons fixed at the vertices determine the spatial arrangement of chemical bonds. In order to eliminate this contradiction, it is necessary, he says, to accept the directions of bonds as normals to the planes of the electron orbits [181].

During that symposium Lewis also tried to explain for the first time how a pair of negatively charged electrons may be formed:

... every atom and every molecule that has an uneven number of electrons possesses a magnetic moment, and the great majority of atoms and molecules which have an even number of electrons possess no magnetic moment (...) It would perhaps be going a little too far to say that it is magnetic force which couples two electronic orbits to form the electron pair (...) For the present it will be sufficient to assert that the coupling of two electronic orbits, with the neutralisation of their magnetic fields and their magnetic moments, is the most fundamental of chemical phenomena [181].

In 1925, the Dutch physicists G. Uhlenbeck and S. Goudsmit [182] showed that the phenomenon of the splitting of spectral lines in a strong magnetic field could be explained if the electrons were assumed to have their own angular momentum whose vector can take a sense in agreement with or opposite to the sense of the orbital vector of angular momentum. This specific angular momentum is called the 'electron's spin'. The spin value of the whole atom can be calculated by the vector addition of the spin values of all electrons in the atom and the spin of its nucleus.

In the same year, even some time before the publication of the Dutch physicists concerning spin, Wolfgang Pauli proposed the hypothesis that two electrons having all the quantum numbers equal cannot occur in one atom. It followed from these two discoveries that two electrons can move along one orbital only when the sense of their spins are opposed [183]. However, this conclusion was formulated two years later by W. Heitler and F. London, who applied quite new mathematical and physical ideas to the solution of the problems of chemical bonds.

These new ideas related to the description of phenomena in the microworld were suggested separately by two German physicists, Erwin Schrödinger and Werner Heisenberg. This gave rise to *quantum mechanics*. Employing the methods of quantum mechanics Heitler and London considered, in 1927, the feasibility of exchange between the nuclei of indistinguishable electrons, and thus developed a quantum theory of valence bonds. In the same year, Friedrich Hund, using quantum equations of atomic orbitals, derived equations which described orbitals comprising whole molecules. Hund in this way laid the foundations of the whole group of methods referred to as methods of molecular orbitals. The works of Heitler, London, and Hund gave rise to a new branch of chemistry—*quantum chemistry*.

Mention should also be made of Paul A. M. Dirac's work published in 1928. After introducing Heisenberg's uncertainty principle to Schrödinger's equation, Dirac showed that particular electrons have their own magnetic moment proportional to their spins, and that the vector of this magnetic moment can be positioned parallel or antiparallel to the direction of the applied magnetic field. Hence, it followed that electrons which revolved along one (atomic or molecular) orbital must have magnetic momenta of opposed senses. In this way he explained why two negative electrons do not repel each other when revolving in one orbital, and thereby Lewis' statement, made five years earlier, was supported.

In 1929, J. S. Slater considered a relationship between the spatial shape of the particular orbitals in an atom and the quantum numbers which describe every orbital. He showed that p orbitals, which are filled only by single electrons in nitrogen and oxygen atoms, form angles of 90° with one another, and the axes of these orbitals determine the directions of the particular chemical bonds. In reality the angles between these bonds are greater because identical atoms (e.g. hydrogen atoms), bonded with the central atom through these orbitals, repel one another. In this way Slater explained the spatial arrangement of atoms in the compounds of nitrogen and oxygen; thus he explained the structure of a molecule of ammonia and its derivatives, as well as the structure of a molecule of water. The quantum foundations of the notion of the tetrahedral shape of the carbon atom were formulated by Linus Pauling who introduced the concept of 'orbital hybridization'

4.10. ELECTRONS AND THE CHEMICAL BOND

in the years 1935–1940. He showed that the linear combination of Slater's orbitals leads to the conclusion that angles between orbitals undergo change. In a carbon atom one s orbital (of spherical symmetry), and three p orbitals are filled by single electrons. The linear combinations of the s orbital with one, two, or three p orbitals give, as a result of hybridization, the following arrangements, respectively: two linearly arranged bonds (sp hybridization), three bonds in a plane directed towards the vertices of a triangle (sp^2 hybridization), or four bonds directed towards the vertices of a tetrahedron (sp^3 hybridization). In the first two arrangements the p orbitals of the neighbouring atoms which do not take part in p hybridization contribute to the formation of multiple bonds.

Thus, it was thanks to quantum chemistry that the theoretical foundations of model stereochemical considerations were formulated and their correctness was justified.

In 1940, N. V. Sidgwick and C. F. Powell suggested that the calculations of hybridized atomic orbitals should also include valency orbitals filled with a coupled electron pair, i.e. a pair which cannot form a normal chemical bond. In consequence, it was possible to explain by the methods of quantum chemistry the formation of not only valence bonds, which corresponded to the main valency numbers, but the so-called 'semi-polar' or 'coordinate bonds' as well; they are formed when two atoms share a pair of electrons which initially belonged to either of them. The proposal of Sidgwick and Powell also allows an explanation of the notion of secondary valency introduced by Werner in 1891 in order to account for the structure of complex compounds.

Today we know that every atomic or molecular system owes its stability to a definite spatial arrangement of negative electrons around the relatively immobile positive nuclei or, more precisely, atomic kernels. It follows from quantum chemistry that the formation of a new bond between initially separate systems of positive and negative charges (of atoms or molecules) consists in changing the arrangement of the negative electrons both in the area between overlapping systems and within them. Such an approach to the mechanism of the formation of a chemical bond involves the formation of ionic valency, and coordinate bonds as well as intermolecular interactions. Thus, Werner's view is confirmed that there is no significant difference between primary and secondary valency.

Attempts were also made to calculate by quantum methods the probable structure of molecules. In 1940, Linus Pauling developed a 'theory of chemical resonance'. As a starting point this assumes different possible structures of a compound. It reminds us of Kekule's approach; by a mathematical combination of wave functions corresponding to those structures it permits an energetic description of the most useful structure of the isolated molecule, which thereby fulfils Butlerov's dream. An improper approach to the resonance structures as transitory forms once provoked a violent criticism of this theory in the Soviet Union. On the basis of the resonance theory one can *ex post* explain the observed phenomena but it is scarcely possible to predict them.

The development of computers has made it possible to perform more accurate calculations of the distribution of electron density, even in systems involving several molecules. Thanks to the methods of quantum chemistry we are today able to understand changes in the molecular structure under the influence of the surrounding molecules. We realize that the structure of a molecule depends on its surrounding as well as on molecular interactions. Modern mathematical methods permit us not only to calculate but also to predict the ability of molecules of known structures to enter into reactions.

The considerations on the structure of molecules were initially associated with the affinity of atoms and their groups. In some periods, i.e. in the second half of the 19th century and in the first half of the 20th century, structure was a separate area of investigation. Now, thanks to the development of new methods, we can once again—in a more advanced manner—consider simultaneously the structure of a chemical substance and its ability to react.

CHAPTER 5

Capacity of a Substance for Transformation

5.1. Antiquity

Even a cursory observation of the transformations of substances enables one to conclude that some transformations occur more rapidly and more easily, whereas others occur slowly and with a greater difficulty. Transformations were observed to occur in animate nature, e.g. the growth of plants and the reproduction of animals, and also in inanimate nature, e.g. during metallurgical processes. As we have said earlier, in Chapter 1, the union of contraries was believed to be the factor that caused substances to be transformed. In the thinking of primitive peoples as well as ancient and mediaeval philosophers this element played an essential role. On the other hand, it was also observed that transformation occurred readily when the substrate was only little different from the product. In the treatise 'On Generation and Corruption' Aristotle analyses factors that favour transformations, i.e. contrarieties, similarity, and also dispersion:

> Thus it is clear that only those agents are 'combinable' which involve a contrariety—for these are such as to suffer action reciprocally. And, further, they combine more freely if small pieces of each of them are juxtaposed. For in that condition they change one another more easily and more quickly; whereas this effect takes a long time when agent and patient are present in bulk (...) For instance, liquids are the most 'combinable' of all bodies [23, 328a].

However, not all bodies can be combined. A problem thus arose: which bodies can form a mixture, and what happens to them after being mixed and 'combined'. And it must be remembered, as we have underlined in the foregoing chapters, that the ancient scientists drew a distinction between homogeneous and heterogeneous mixtures, but they failed to distinguish the former from chemical compounds.

Heterogeneous mixtures were regarded as 'compositions', and homogeneous mixtures (compounds) as 'combinations'.

Aristotle writes on the ability of substances to combine:

> It is clear, then, from the foregoing account, that 'combination' occurs, what it is, to what it is due, and what kind of thing is 'combinable'. The phenomenon depends upon the fact that some things are such as to be (a) reciprocally susceptible and (b) readily adaptable in shape, i.e. easily divisible. For such things can be 'combined' without its being necessary either that they should have been destroyed or that they should survive absolutely unaltered: and their 'combination' need not be a 'composition', nor merely 'relative to perception'. On the contrary: anything is 'combinable' which, being readily adaptable in shape, is such as to suffer action and to act; and it is 'combinable' with another thing similarly characterized (for the 'combinable' is relative to the 'combinable'); and 'combination' is unification of the 'combinables', resulting from their 'alteration' [23, 328b].

The display of contrary and similar qualities at the same time by the constituents of a mixture or a combination now seems to us contradictory. However, some Greek philosophers took both factors into consideration. Aristotle recounts their views:

> For (i) most thinkers are unanimous in maintaining (a) that like is always unaffected by 'like', because (as they argue) neither of two 'likes' is more apt than the other either to act or to suffer action (...) and (b) that 'unlikes' i.e. differents are by nature such as to act and suffer action reciprocally. (...).
>
> But (ii) Democritus dissented from all the other thinkers and maintained a theory peculiar to himself. He asserts that agent and patient are identical, i.e. 'like'. It is not possible (he says) that 'others' i.e. 'differents' should suffer action from one another; on the contrary, even if two things being 'others', do act in some way on one another, this happens to them not *qua* 'others' but *qua* possessing an identical property [23, 323b].

Considering the possibility of reciprocal transformation of the elements, Aristotle realized that it may be due to the existence of a necessary contrariety between them, but—at the same time—he presumed that too great a contrariety might not facilitate transformation. He made it explicit in the following fragment:

> (...) the 'elements' all involve a contrariety in their mutual relations because their distinctive qualities are contrary. For in some of them both qualities are contrary—e.g. in Fire and Water, (...) while in others one of the qualities (though only one) is contrary—e.g. in Air and Water, the first being moist and

hot, and the second moist and cold. It is evident, therefore, if we consider them in general, that every one is by nature such as to come-to-be out of every one: and when we come to consider them severally, it is not difficult to see the manner in which their transformation is effected. For, though all will result from all, both the speed and the facility of their conversion will differ in degree. Thus (i) the process of conversion will be quick between those which have interchangeable 'complementary factors', but slow between those which have none. The reason is that it is easier for a single thing to change than for many [23, 331a].

Thus, we can see that according to Aristotle water can easily and quickly be transformed into air, whereas transformation of earth into air must be much slower. Thus for Aristotle the rate of transformation was a proof of the ease of transformation; it was due to the two opposing agents; the contrariety of properties and their similarity (affinity). In his other works, the Greek philosopher was also concerned with the effective cause of such transformations—their 'driving force'. As we have already mentioned, Aristotle distinguished matter and form. In his view matter is a potential being realized by form. Transformations occur when one form disappears and another appears. The agent which originates such transformations is referred to as a causal determinant. It acts by passing from the state of possibility to the state of accomplishment. One of the translators of Aristotle's works, L. Regner points out that what acts as a causal determinant must be—according to Aristotle—in the state of accomplishment. The state of accomplishment or realization of the causal determinant is referred to by the Greek philosopher as *entelechia* or *energeia*. The opposite notion to 'accomplishment' is 'possibility of existence'—*dynamis*.

In Aristotle's 'Physics', the following consideration can be found:

The fulfillment (*entelechia*) of what exists potentially, in so far as it exists potentially, is motion—namely, of what is alterable *qua* alterable, alteration of what can be increased and its opposite what can be decreased (there is no common name), increase and decrease: of what can come to be and can pass away, coming to be and passing away: of what can be carried along, locomotion [23, 201a].

On the other hand, in the treatise 'On Generation and Corruption' Aristotle applies the concept of actuality (*energeia, entelechia*) to the very existence of a body. He writes: "(...) every perceptible body should be indivisible as well as divisible at any and every point. For the second predicate will attach to it potentially, but the first actually" [23, 316b], or elsewhere he says: "(...) some things are-potentially while others are-actually" [23, 327b]. In the treatise 'On the Soul', the philosopher states that "the Soul is the primary entelecheia of a natural body (...). It must be, however, an organic body". Next he recalls the concepts of matter and form and concludes his considerations that "matter is potentiality, form is actuality (entelechia)" [23, 412b].

We have devoted much space to the Aristotelian concept of entelechia because this obscure concept seems to have played an important role in subsequent conceptions accounting for the formation of organic substances.

In the considerations discussed above, Aristotle aimed at defining the conditions which determine whether change can occur, and whether it occurs easily or not. Aristotle was also concerned with another aspect of change which is related to the role of fire. We have seen that the Greek philosophers treated fire as one of the four elements which carried heat and dryness. We have mentioned that the remaining three elements—earth, water, and air—in fact represented the three states of aggregation. Only fire had no particular 'assignment'. Greek philosophers were quite aware of the distinctive role of fire. They held that the elements interact with one another but fire is the only element which acts on other bodies without being acted upon by these bodies. Aristotle frequently mentioned the effect of qualities on different bodies; however, qualities do not act upon one another. As we have seen, he considered heat and cold as active qualities, whereas dryness and moisture were passive qualities. However, this does not refer to the elements which carry these qualities. Except for fire which, as we have seen in Chapter 2, is the determinant of changes "(...) for 'dissociating', which people attribute to Fire as its function, is associating things of the same class, since its effect is to eliminate what is foreign" [23, 329b].

Fire was therefore regarded as an element by the ancient philosophers, but at the same time it has features which characterize its basic qualities. Comparing the process, or rather the mechanism of different transformations, the ancient Greeks thus considered two aspects: the rate or ease of transformations and the agents which provoke them. These two problems connected with the mechanism of reaction, will reappear continuously throughout the whole development of chemistry.

5.2. Chemistry in the 11th–18th Centuries

In Chapter 2, while discussing the development of the concept 'element', we dealt with the significance of the spagyric elements, one of them being sulphur—the carrier of combustibility. Sulphur used to be closely associated in the minds of scientists with fire. Therefore, this spagyric element—sulphur was often identified with the peripatetic element—fire, one of the chief factors which brought about the transformations of bodies. Mediaeval chemists were convinced that for a new body to appear, two bodies of opposite properties must be united. In Chapter 1, we have shown that opposite properties were conceived of as differences of the sexes. We have quoted a fragment of Maier's treatise from 1616 which illustrates this approach. In the *Hermetico-Spagyrisches Lustgärtlein*, published ten years later, we find an alchemical emblem of the 'most eminent of the philosophers' (Latin: *Palmarius inter philosophi*) with the following maxim: "Match a flunkey with an attractive sister and they will bear a son unlike the parents" [46]. This maxim refers to a chemical reaction, of course.

Early in the 18th century, in his lectures to students, the Dutch physician and chemist Herman Boerhaave characterized the views on the formation of bodies which prevailed in alchemy in the following way:

I come now to add a few, but candid and ingenuous considerations, on the great use of chemistry in alchemy. To speak my mind freely, I have not met any writers on natural philosophy, who tread (*sic*) of the nature of bodies, and the manner of changing them, so profoundly, or explain'd them so clearly, as those called alchemists. (...) Hence they are contitnually inculcating, that man cannot by any art go beyond the powers impressed by the Creator on bodies: of which powers such as are of necessary use to life, are obvious enough; but others lie less apparent, and are only revealed to those who seek them with great labour and industry, but both are equally natural. (...) That wise men observe the works of nature as they offer themselves, and then by experiments endeavour to learn the laws; which the Creator has impressed on his work,

(...) the principal of which laws is, that all things arise from other similar preexistent ones; plants from plants, animals from animals, and fossils from fossils. That all power of propagating is contained in the seminal matter alone; which converts every crude thing it takes, into its own form, and assimilates it to itself. That to have an offspring from this semen always requires a male father, and female mother; so that nothing is ever produced without the natural copulation of these two. They add also that metals alone, by reason of their extreme simplicity, admit of being produced in the shortest time, for heavy mercurial fluid, and a fixing seminal sulphurous power, intimately mixed by the force of fire, and thus united together in an indissoluble bond, so that mercury, or argentum vivum, does the office of mother, and sal vivus, that of father; (...) so where-ever I understand the alchemists, I find them describe the truth in the most simple and nacked terms, without deceiving us, or being deceived themselves [184].

Fig. 17. Distillation laboratory from the 16th century, acc. to a drawing by Jan van den Straet in [7].

Regardless of the rate of transformation, problems related to the term 'affinity' were also considered in the period of alchemy. Albertus Magnus holds that "sulphur will blacken silver and burn metals because of affinity to these bodies" [185]. In the 16th century, Bernard Palissy considers 'affinity' existing between different salts. In the 17th century, Franciscus de la Boe Sylvius explains that precipitation of one metal by another from an acid solution is due to a greater affinity of acid to the latter metal than to the metal which was previously dissolved in acid. In the same period, Johann Rudolf Glauber explains the course of some chemical reactions by a higher affinity of one of the compounded substances to the substance added than to the substance with which it previously formed a compound. In this way he explains the precipitation of gold previously dissolved in *aqua regia* when 'silicate liquor' (potassium silicate) is added to the solution. He also states that ammonia is precipitated from salammoniac heated with zinc oxide because the "latter prefers acids and is preferred by them too". Further, he notes that sulphuric acid has a greater 'love' for sodium carbonate than nitric acid. Note that Glauber does not employ the term 'affinity', but in accordance with the alchemical approach he speaks of the 'love' of some substances to others.

5.3. Boyle's Corpuscular Approach

The view on the mechanism of the combining and mixing of substances underwent transformation throughout the 17th century due to the spread of the corpuscular conception of the texture of matter. The reasoning of the contemporary scientist is illustrated by considerations expressed by Carneades in Boyle's 'The Sceptical Chymist':

> I consider that it very often happens that the small parts of bodies cohere together but by immediate contact and rest (...) but that it is possible to meet with some other body, whose small parts may get between them, and so disjoyn them; or may be fitted to cohere more strongly with some of them, than those some do with the rest; or at least may be combined so closely with them, as that neither the fire, nor the usual instruments of chymical anatomies will separate them.

Boyle does not, however, rule out the fact that mixtures and alloys exist whose properties are different from those of their constituents and that such constituents can be recovered in their original form. As an example he gives an alloy of gold and silver. Further, Carneades goes on to state:

> But (...) there are other clusters wherein the particles stick not so close together, but that they may meet with corpuscles of another denomination, which are disposed to be more closely united with some of them, than they were among themselves. And in such case, two thus combining corpuscles losing that shape, or size, or motion, or other accident, (...) each of them really ceases to be a corpuscle of the same denomination it was before; and from the coalition of these there may emerge a new body as really one, as either of the corpuscles was before they were mingled (...) since this concretion (...) can no more by the fire or any other known way of analysis, be divided again into the corpuscles that at first concurred to make it, than either of them could by the same means be subdivided into other particles [32, pp. 87–88].

Thus, on the basis of the corpuscular theory, Boyle distinguished a mixture and a compound, although he expressed it in a rather unclear way so, as a matter of fact, his contemporaries did not notice that distinction. By means of the corpuscular theory Boyle also attempted to account for the fact that only some substances can react with one another by particle cohesion. These considerations concern the problem of how easy it is for a substance to react.

In 'The Sceptical Chymist' we also find another kind of consideration which is related to the role of fire during chemical reactions. The author refutes the Aristotelian concept of the effect of fire, which he formulated as: *congregare homogenae et heterogenae segregare* (to combine the homogeneous and to separate the heterogeneous); he shows that cold can also have a similar property of combining bodies, e.g. upon freezing, wine gives off crystals of solidified water. Furthermore, Boyle states that the effect of fire on the same substance may be different, and that "the same portion of matter may easily by the operation of the fire be turned (...) into the form of a brittle and transparent, or an opaceous and malleable body" [32, p. 217]. Boyle writes:

> (...) the true and genuine property of heat is, to set a moving, and thereby to dissociate the parts of bodies, and subdivide them into minute particles, without regard to their being homogeneous or heterogeneous, as is apparent in the boyling of water, the distillation of quicksilver, or the exposing of

bodies to the action of the fire (...) where, all that the fire can do, is to divide the body into very minute parts which are of the same nature (...) And even when the fire seems most so *congregare homogenea, et segregare heterogenea*, it produces that effect but by accident [32, p. 54].

Moreover, the fire sometimes does not separate, so much as unite, bodies of a differing nature; provided they be of an almost resembling fixedness, and have in the figure of their parts an aptness to coalition, as we see in the making of many plaisters, oyntments, etc. [32, pp. 56–57].

Next Boyle defines more explicitly the role of fire in connection with the corpuscular approach to the texture of matter.

(...) the operation of the fire does actually (sometimes) not only divide compounded bodies into small parts, but compound those parts after a new manner, whence consequently for ought we know, there may emerge as well saline and sulphureous substances, as bodies of other textures [32, p. 76].

Thus Boyle, like the ancient Greeks, tried to imagine why some transformations do take place and others do not; and he also analysed what factors can cause these transformations to occur. He attributed a great role to fire with surprising acuity for his time. Boyle described the intensification of corpuscular motion under the effect of fire.

5.4. Affinity in the 18th Century

We have already mentioned that Boyle's work, which contained many revolutionary ideas, exerted a very slow influence on the development of chemical concepts. It contained first of all a critique of the old views, and the new theories, which were to replace the old ones rejected by the author, were put forward without the necessary emphasis. As we have seen in Chapter 2, at the turn of the 17th century the spagyric sulphur, i.e. the peripatetic fire, was absorbed by Becher's concept of 'fatty earth', Lemery's concept of oil, and subsequently by Stahl's theory of phlogiston—the element of combustibility. The presence of this element of combustibility in a body governed the possibility of the reaction which we now call oxidation.

Throughout the 18th century some scientists identified the matter of fire with the matter of heat, also known as caloric; others, e.g. Hermbstädt (cf. Section 2.4) drew a distinction between the two matters. Phlogiston was thus believed to be an agent which makes a chemical reaction possible. Today we would compare it to potential chemical energy. Caloric, on the other hand, which according to the prevailing views of Lavoisier and Dalton opposed the attraction of particles, might be compared to the kinetic energy of the particles of a substance. Thus, in the period when the concept of energy was not well formulated, phlogiston and caloric played the role of this very concept, and their use permitted, later on, the retirement of the concept of energy.

In the early 18th century, Newton attempted to account for the problem of the association of bodies. In Query 31 of the third part of the second edition of his 'Opticks' the English physicist writes:

Have not the small particles of bodies certain powers, or forces, by which they act a distance, not only upon the rays of light for reflecting, refracting, and inflecting them, but also upon one another for producing a great part of the phenomena of Nature? For it's well known that bodies act one upon another by the attactions of gravity, magnetism, and electricity; and these instances shew the tenor and course of Nature, and make it not improbable but that there may be more attractive powers than these. (...) How these attractions may be performed I do not here consider. What I call attraction may be performed by impulse, or by some other means unknown to me. I use that word here to signify only in general any force by which bodies tend towards one another, whatsoever be the cause [79, p. 376].

These forces of attraction between particular substances cannot—according to Newton—always have the same value because:

When spirit of vitriol (sulphuric acid—R.M.) poured upon common salt or saltpetre makes an ebullition with the salt, and unites with it, and in distillation the spirit of the common salt (hydrochloric acid—R.M.) or saltpetre (nitric acid—R.M.) comes over much easier than it would do before, and the acid part of the spirit of vitriol stays behind, does not this argue that the fixed alkali of the salt attracts the acid spirit of the vitriol more strongly than its own spirit, and not being able to hold them both, lets go its own? [79, p. 378].

Newton accounts for many properties of substances by the differences in the degree of attraction, including the sense of taste by which flavour is known. He goes on to say:

Do not the sharp and pungent tastes of acids arise from the strong attraction whereby the acid particles rush upon and agitate the particles of the tongue?

5.4. AFFINITY IN THE 18TH CENTURY

And when metals are dissolved in acid menstruums, and the acids in conjunction with the metal act after a different manner, so that the compound has a different taste much milder than before, and sometimes a sweet one—is it not because the acids adhere to the metallic particles, and thereby lose much of their activity? [79, p. 386].

Along with Newton's considerations and the modifications of the concept of an agent which brought about chemical reactions, the concept of the tendency of different bodies to unite, or their affinity, became more and more concrete in the 18th century. In 1718, the French pharmacist E. F. Geoffroy collected observations concerning the different degrees of an attractive force between substances that causes them to enter into a chemical combination. In a memoir presented at the Academy of Sciences Paris, he wrote:

We observe in chemistry certain affinities between different bodies according to which they unite easily, one with another. These affinities have their degrees and are governed by their laws. We observe their different degrees when, among many mixed materials which have some disposition to unite, we note that one of the substances invariably unites with a certain other one in preference to all the rest. (...) Whenever two substances which have some disposition to unite, the one with the other, are united together and a third which has more affinity for one of the two is added, the third will unite with one of these, separating it from the other. (...) I show today in this table the different affinities which I have collected from the observation of different chemists as much as from my own. (...) The first line of this table includes different substances used in chemistry. Below each substance different types of materials are arranged in columns in the order of their affinity for that substance such that that which is nearest has the greatest affinity for the substance and cannot be displaced by any of the materials below it, but that it may remove any of the lower ones when they are joined to the substance [186].

Thus, Geoffroy approached the problem of affinity in a semi-quantitative way. He did not at that time know of any measure of affinity, but he tried to arrange reactions according to the degree with which the substances united easily with one another. His table (cf. Fig. 18) concerns, of course, both chemical reactions (e.g. between acids and basic substances) and physical processes of dissolution (e.g. amalgamation of metals by mercury, mixing of water and alcohol, dissolution of salt in water). This table was used by many 18th-century chemists; a hundred years later, it was mentioned by Jędrzej Śniadecki in the third edition of his textbook 'The Principles of Chemistry'.

5. CAPACITY OF A SUBSTANCE

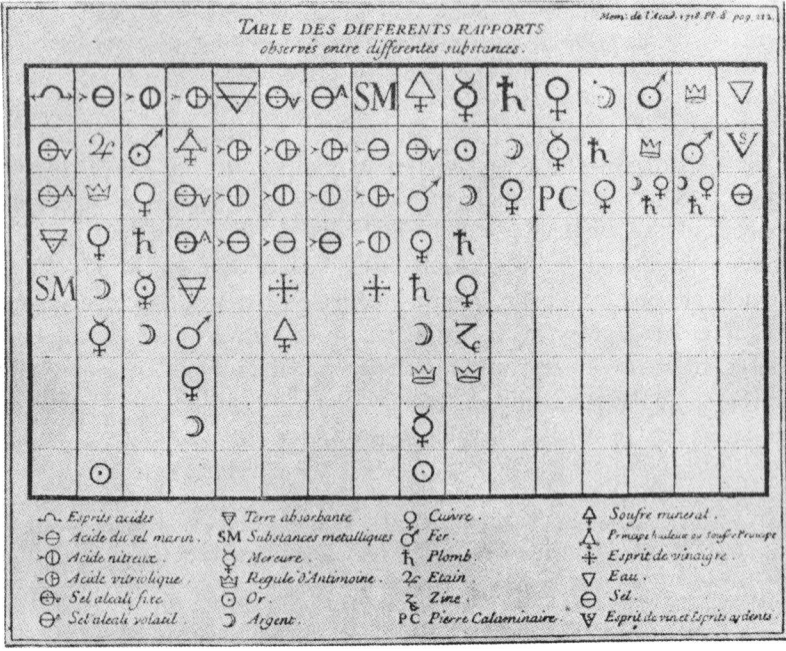

Fig. 18. Geoffroy's table of affinities, acc. to [9].

Symbols represented in the lower part of the table in sequence corresponding to their occurrence in the columns have the following meanings (a literal translation is provided, and in brackets—if possible—the present name of a substance):

1. Spirit of acids (volatile acid oxides)
2. Acid of marine salt (hydrochloric acid)
3. Saltpetre acid (nitric acid)
4. Vitriol acid (sulphuric acid)
5. Solid alkalic salt (potash, soda)
6. Volatile alkalic salt (ammonia, salammoniac)
7. Absorbing earth
8. Metallic substances
9. Mercury
10. King of Antimony (metallic grains of antimony)
11. Gold
12. Silver
13. Copper
14. Iron
15. Lead
16. Tin
17. Zinc
18. Calamine stone (silicates)
19. Mineral sulphur (mined sulphur)
20. Fatty principle or sulphur principle
21. Acetic spirit (acetic acid)
22. Water
23. Salt
24. Vinous spirit and inflammable spirits (alcohols)

5.4. AFFINITY IN THE 18TH CENTURY

The next step in the development of the concept of affinity is to be found in the works of the Swedish chemist Thorbjön Bergman. He analyses the Newtonian 'attraction' between different chemical substances, and he mentions that what he calls 'attraction' is referred to as 'affinity' by others. In 1775, Bergman published a work entitled *De attractionibus electivis* in which he writes:

Several species of contiguous attraction may be distinguished. I shall here briefly mention the principal. When homogeneous bodies tend to union, an increase of mass only takes place, the nature of the body remaining still the same, and this effect is denominated attraction of aggregation. But heterogeneous substances when mixed together, and left to themselves to form combinations are influenced by difference of quality rather than of quantity. This we call attraction of composition (...). When it takes place between three substances respectively, to the exclusion of one, it is said to be single elective attraction; when between two compounds, each consisting of only two proximate principles which are exchanged in consequence mixture, it is intituled double attraction [187, p. 291].

'Attraction', as distinguished by Bergman, is denoted by many terms even today. We speak of reactions of synthesis, and of reactions of single and double exchange, which precisely correspond to the kinds of Bergman's attraction.

Bergman was aware that attraction/affinity can vary under the influence of temperature. He wrote:

The only external condition, which either weakens or totally inverts the affinities of bodies subjected to experiments, is the different intensity of heat. But this cause can only operate in cases where the same degree of heat renders some bodies remarkably volatile in comparison to others. Suppose A to be attracted by two other substances; and let the more powerful act at the ordinary temperature with the force a, the weaker with the force b: suppose, at the same time, the former to be the more volatile; let its effort to arise be expressed by V, and that of the other by v. When these three substances are mixed together, the stronger will attract A with a force $= a-b$: but should the heat be gradually raised, this superior force will be more and more diminished; and as V will increase faster than v, we shall at last have $a-b = V-v$. This state of equilibrium will be immediately destroyed by the smallest addition of heat; and thus b, which was before the weaker, and incapable of producing any effect, will now prevail [187, p. 299].

In the above fragment, we find for the first time a faint attempt at a mathematical formulation of the problem of affinity. Furthermore, also for the first time, Bergman showed that raised temperature may change affinities, or—we should say today—reverse the direction of a chemical reaction.

5.5. Heat as a Measure of Affinity

A distinctly quantitative approach to affinity will be found in *Memoire sur la chaleur* published by Antoine Lavoisier and Pierre Simon de Laplace in 1783:

> (...) when combining substances act on one another according to the degree of mutual affinity, thus their molecules being subject to the forces of mutual attraction which can change the amount of live force and thereby the amount of heat [188].

It follows from the above quotation that the French investigators seem to have suggested that emitted heat might be considered as a measure of the mutual affinity of two substances. It was the first concrete attempt at a quantitative determination of affinity which linked two problems: a tendency of substances to combine with one another and the factor which facilitated such a combination.

The term 'live force' (French *force vive*) used in the above quotation is an equivalent of the present-day term 'energy' which was not used in the 18th century. Throughout the 18th century scientists made a distinction between, so to say, a 'dead force', which played a role in statics, i.e. in the branch of physics dealing with the conditions of mechanical equilibrium, and a 'live force' connected with the dynamics of motion. The problem posed by Galileo, whether the measure of the rectilinear motion of a body is a quantity proportional to its velocity or a quantity proportional to the square of its velocity, was solved in the mid-18th century by the French physicist Jean Le Rond d'Alembert. He distinguished a 'quantity of motion' (today called 'momentum'), which equals the product of mass and velocity, and a quantity of work, which a moving body can do, and which is proportional to the product of mass and the square of velocity. He called the second quantity a 'live force'. It was in 1807 that Thomas Young proved that work done equals half of that product, and he

5.5. HEAT AS A MEASURE OF AFFINITY

called this quantity 'energy'. However, the term 'energy' had been used by chemists before; in his works from the years 1801 and 1803, Claude Louis Berthollet compared values of the energy of mutual affinity of different compounds.

However, in the early 19th century heat was not identified with energy. In Chapter 2, we have presented reasons why the substantive theory of heat was still used for another fifty years or so, in spite of the quite convincing arguments given by Rumford. Admittedly, the considerations of Lavoisier and Laplace enabled the problems of affinity to be approached quantitatively; however, the very concept of affinity was still obscure. The works of Lavoisier's friend and collaborator, Berthollet, mentioned earlier, made a significant contribution to the development of this concept. In a publication dating from 1801, he describes the action of two acids on one salt or base and the action of one acid on two salts or bases, and on the basis of his observations he concludes:

It results then from the preceding experiments (...) that when a substance acts on a combination, the subject of combination divides itself between the two others, not only in proportion to the energy of their respective affinities, but also in proportion to their quantities.

Thus Berthollet is the first to point out that a chemical reaction is due not only to the affinity of reacting substances but also to their amounts.

I have considered all the forces which can, by their concurrence with, or opposition to, the reciprocal affinity of substances acting according to the preceding principle, exert any influence on chemical combinations and phenomena. They are reducible to the following: the action of solvents; or the affinity which they exert in proportion to their quantity; the force of cohesion, which is the effect of the mutual affinity of the parts of a simple or compound substance; elasticity, whether natural or produced by caloric, which ought to be considered as an effect of the affinity of caloric; efflorescence, which may be attributed to an affinity not yet determined, acts only in very rare circumstances; gravitation too exerts some influence, particularly when it produces the compression of elastic fluids; but no inconvenience can result from its being confounded with the force of cohesion.

I have endeavoured to find if it were possible to ascertain the relative affinity of two substances by means of a third; and I have observed that, in order to do so, it would be necessary to ascertain in what proportion that third substance would combine with a given quantity of the two former (...). The very term, elective affinity must lead into error, as it supposes the union of the whole of one substance with another, in preference to the union of the whole of one substance with another, in preference to the union of the whole of one substance with another, in preference to a third; whereas there is only a partition of action, which is itself subordinate to other chemical circumstances [189].

In his next and fundamental work, published in 1803, entitled *Essai de statique chimique*, Berthollet analyses the terms affinity and chemical attraction. In the Introduction he emphasizes the significance of both terms:

All forces which produce chemical phenomena come from the mutual attraction of the molecules of bodies which is referred to as affinity in order to distinguish it from astronomical attraction. It is possible that both have the same property, but astronomical attraction is exerted only between masses over distance, where the form of molecules, their distances and their particular susceptibility have no influence; their effects being always proportional to mass and inversely proportional to the square of distance can be precisely calculated. The effects of chemical attraction or affinity are, on the other hand, so much altered by particular conditions and frequently indefinite that it is impossible to derive one general principle, but they must be gradually ascertained (...) since it is very likely that in its origin affinity does not differ generally from attraction it should also be expressed by the laws of mechanics and included in the phenomena which depend on the action of mass, (because) every substance which tends to enter into combination, acts proportionally to its affinity and its quantity.

Chemical action of a substance depends not only on affinity, which is characteristic of its constituent parts, and on quantity; it also depends on the state in which these constituents are (i.e. on chemical and physical conditions—R.M.) (...) these conditions, by changing the properties of elementary particles of a substance, form what I shall call constitution; it is necessary to know not only each of these conditions, but also the circumstances upon which they depend [86, Vol. I, p. 5].

The following statement, contained in the work quoted, shows how much importance Berthollet attributes to affinity:

Chemistry may be regarded as a science which has general principles only from the moment when affinity was recognized as the cause of all combinations (i.e. compounds and mixtures—R.M.).

5.5. HEAT AS A MEASURE OF AFFINITY

As can be seen from the above quotations, at the turn of the 18th century affinity was closely linked with attraction. The great role attributed to attraction in natural science was due to the influence of Newton's ideas, which were then widespread. Although in *Memoir sur la chaleur*, Lavoisier and Laplace suggested that liberated heat be considered as a measure of affinity in his textbook *Traité élémentaire de chimie* Lavoisier himself contrasted heat with attraction, as we have shown in Chapter 2. A similar view was held as we have seen, by Dalton, who also considered "affinity maintained in equilibrium by the repulsion force produced by caloric" [87, Vol. I, p. 182].

Berzelius devotes to affinity one of the first chapters of his textbook where he includes the considerations of both Bergman and Berthollet:

The whole world which surrounds us, with all compound bodies infinitely different from one another of which it is formed, consists of a small number of simple substances which thanks to specific forces called affinity combine with one another in different ratios.

These affinities are of a twofold kind. One force thanks to which small body particles are combined is referred to as affinity of union (German *Zusammenhangsverwandschaft*) or force of cohesion. Different stability of these bodies depends on different degree of this affinity; the body is hard and solid if this affinity is high, the body is liquid like water if it is lower, and when it diminishes, the body turns into air or gas.

Berzelius also attributes the formation of crystals to the effect of affinity. He writes further:

Another kind of affinity is called affinity of combination in homogeneous bodies, e.g. between the tiniest grains of cinnabar—it is a chemical affinity or affinity of combination, but it also occurs between heterogeneous bodies, e.g. between sulphur and mercury.

Affinity of combination exhibits different variations of which the following are most important:

(1) The degrees of this affinity for majority of bodies which are affined through it are never equal, but in one body it is always higher than in another. Thus, for example, iron has a higher affinity for sulphur than for mercury; therefore when iron swarves are sufficiently mixed with cinnabar and heated together, then iron will combine with sulphur and mercury will be given off. This change of affinity is called affinity of choice.

(2) The second difference is shown with respect of the number of bodies which affect on one another through affinity, namely it appears that a greater

part of a body with weaker affinity of combination prevails over the smaller part of a second body with stronger affinity; in other words: sometimes quantity can substitute for lack of force [76, Vol. I, p. 31].

Thus, in the first half of the last century, while searching for the cause of chemical reactions, the investigators explained the degree of easiness of the formation of chemical compounds and solutions by a rather vague phenomenological concept of chemical affinity, although then it was not distinguished from the rate of reactions. This finds evidence in the following fragment from Berthollet's work:

Chemical action occurs at a faster or slower rate, and this situation has often an influence on the results (...). Compounds which seem to be stable in their proportions are decomposed by a slower action than that which forms them (...). Chemical action is slower when it is weak than when it is strong, and as the action of a given substance becomes weaker together with its saturation (i.e. neutralization—R.M.), then the final stages of saturation take much longer time than that which is required to approach it [86, Vol. I, p. 409].

We shall refer to the above fragment again in this chapter when the problem of equilibrium and reaction rate is discussed in more detail. Now we would like only to draw the reader's attention to a semi-quantitative link between the reaction rate and the affinity energy, because the affinity rate had not been determined quantitatively by then.

Among different attempts at a systematization of the elements, the phenomenological affinity was sometimes regarded as one of the criteria of systematization, which has been discussed in Chapter 3.

5.6. Formation of Substances in Living Organisms

Earlier we have discussed problems related to the formation of new substances in the mineral world. From the 17th century views were also developed concerning the formation of substances by living organisms. Of course, those views were based on the conceptions derived from ancient and mediaeval times, and particularly on Aristotle's entelechia. From the Middle Ages came the popular conception of semen providing basis for development of similar entities, which was emphasized in Boerhaave's lecture. Thus, there

5.6. SUBSTANCES IN LIVING ORGANISM

was a sharp boundary between the kingdom of minerals and those of plants and animals. In the first half of the 17th century, the Dutch physician Jan Baptist van Helmont, referring to Aristotle's entelechia, named the agent causing the formation of different substances in plant and animal organisms a 'vital force' (Latin *vis vitalis*). This theory lasted in various forms and modifications until the beginning of the present century. Jędrzej Śniadecki based some of his views on this theory and he developed it in his chemical textbooks ('The Principles of Chemistry') and in a work published in 1804: *Teorya jestestw organicznych* ('A Theory of Organic Entities'). In the third edition of 'The Principles of Chemistry', he points out the necessary existence of such a force:

Looking upon Nature we see that it is rich in an infinite multitude of entities which our art can never create or imitate, which must lie beyond the fringe of our skills (...). Thus the very first remark about them is that they are not simple chemical forms (...). What are organic entities (...) when we look closer upon them, we see that they conceive, grow, (...) and eventually die. Reflecting upon such entities with attention, we see that they are never created in another way but one out of another. Hence we conclude that (...) the power of creation of organic entities, i.e. the power of organization is attributed to organic entities alone (...) experience shows us that the annihilation of organization without the least damage to matter causes this power to disappear immediately; therefore we conclude that the organizing power is not a property of matter but of organization. For this reason it should be called an organic power.

Further, Śniadecki asserts that all matter is

(...) endowed by Nature with certain properties and forces which make it operative (...). Such powers being inseparable from matter include first of all affinities. Thus matter being constituent of organic entities acts upon itself through affinities; and since organization acts upon it simultaneously through organic power, the forces of affinity must acquire a different direction in such entities [51, Vol. II, p. 4].

As late as 1804, Śniadecki emphasizes the unity of all matter:

Today when chemistry can boast of the most perfect separation of organic entities, we know that the constituent elements of all organic entities are the same and their number is very limited. The whole kingdom of plants, as it lives thanks to water and carbonic acid, may be decomposed into water, carbonic acid, i.e. carbon, hydrogen, and oxygen. All animals are composed of these same elements and nitrogen (...). To these elements we can include phosphorus and sulphur in both the plant and animal worlds, albeit these elements being represented less amply in organic entities [190, p. 157].

Similarly, in the twelfth chapter of 'The Theory of Organic Entities' Śniadecki writes:

(...) while considering the very origin of things we have concluded that an elemental force must have been exerted on matter, which then formed the presently living kinds and species, and this force has been referred to as organizing or organic (...). Each (...) matter being transformed into an organic entity is derived to a greater or lesser extent from its chemical existence, and therefore its static affinities and physical compound must be based on such existence, too. And since the power of the matter of heat is continuously exerted against bonds and static affinities it is helpful to the organizing forces in assimilating sustainable matter and, furthermore, it supports and facilitates the action of forces which break up chemical combinations (...).

Due to such action of organic, chemical forces, and of heat it turned out first that since two processes, i.e. organizing and disintegrating ones take place in every living entity, in the former the organic forces are predominant and superior, and disengage matter from physical and chemical laws more or less; in the latter they gradually lose their strength and they respectively yield to chemical forces and allow to be superseded. For this reason I sometimes call the latter process a chemical one. Secondly, since the combination and union of matter from organic entities can never be regarded as simply a chemical one, because in every case it is the result of mutual attachment and a certain balance of organizing and chemical forces, between which the organic sophistication of matter is the greater the more the latter forces are suppressed and conversely [190, p. 205].

As can be seen Śniadecki, having recognized the existence of an 'organizing force', realized that organic substances are made up of the same elements as mineral substances, but they are 'organized' in a different manner. He also understood that organic matter is subject to the same chemical laws as inorganic matter, although he believed that there are also some specific 'organic' laws. We shall discuss this problem later on.

As we have said in the preceding chapter, from the end of the 18th century more and more substances which we now consider to be organic compounds were synthesized. They were obtained from

5.6. SUBSTANCES IN LIVING ORGANISM

mineral substances, and products of those syntheses were not included in the kingdom of plants, nor of animals, either.

It was not until 1828 that Friedrich Wöhler produced urea from ammonium cyanate, which was regarded as a mineral compound, and which by then had only been found in animal urine. It was thus the first substance obtained by the ordinary processes of chemical synthesis which had been previously attributed to the action of 'vital force'.

However, neither Wöhler's synthesis of urea, nor Kolbé's synthesis of acetic acid in 1845, in which the initial material consisted only of inorganic substances: carbon disulfide, chlorine, and amalgam of sodium, could overthrow the concept of vital force.

In his 'Familiar Letters on Chemistry', published for the first time in 1843 and then reprinted many times, Justus Liebig compares the action of both chemical and vital forces:

The cause of the phenomena of life is a force, which does not act at sensible distances; its activity becomes manifest only when the aliments or the blood come into immediate contact with the organ destined their reception, or alteration. The chemical force manifests itself precisely in the same manner; indeed there are no cause in nature producing motion or change in bodies—no powers more closely allied to each other—than the chemical and vital forces [104].

It was quite obvious for Liebig that life processes are accompanied by chemical processes, that it is possible to produce by 'chemical forces' urea, formic acid, oxalic acid, valerian oil and a few other products of vital processes.

These results are enough to justify us in entertaining the hope that we shall, ere long, succeed in producing quinine and morphine, and those combinations of elements of which albumen and fibrine, or muscular fibre, consists, with all their characteristic properties.

Let us, however, carefully distinguish those effects which belong to chemical, from those which depend peculiarly upon vital force, and we shall then be in right channel for obtaining an insight into the later. Chemical action will never be able to produce an eye, a hair, or a leaf [104, p. 24].

On the other hand, Liebig believed that

Neither heat, electricity, nor the vital force, are capable of connecting the particles of two dissimilar elements into a group—of uniting them into a

5. CAPACITY OF A SUBSTANCE

compound; this, the chemical force alone is able to accomplish. (...) The substance of brain, of muscle, the constituents of blood, of milk, of bile, etc., are compound atoms, the formation and duration of which depend upon the affinity which acts between their ultimate particles—their component elements. It is affinity, and no other power, which causes their aggregation. Separated from the living body, withdrawn from the influence of the vital force, it is the chemical forces alone which determine the conditions of their ulterior existence. (...) But Light, Heat, the Vital Force, the Force of Cohesion, and the Force of Gravity, exercise a most decided influence upon the number of simple atoms which unite to form a compound atom, and upon the manner of their arrangement [104, p. 137].

The above text can be found in an unaltered form in a later edition of his letters (1859). However, having recognized the existence of vital force beside chemical force, Liebig criticizes the very use of the term force:

The term 'vital force' in the present state of science, does not denote a force per se, as we may suppose the terms electricity or magnetism to do; but it is a collective term, embracing all those causes on which the vital properties depend. In this sense it is as just, and may by used with as much property, and convey a similar meaning to the term 'force of affinity' or 'chemical force', which denotes the causes of chemical phenomena; of which we know quite as little as we do of them cause or causes which determine vital phenomena [104].

However, as early as 1860, Berthelot tried to show that all organic compounds could be obtained through the action chemical forces. He published a textbook of chemistry of a very significant title *Chimie organique fondé sur la synthèse*. In conclusion he writes:

(...) we have come from the elements, i.e. carbon, hydrogen, oxygen, and nitrogen. Primary double compounds, mainly hydrocarbons, have been created from these elements and only by means of mineral forces. Next we have constructed, on the basis of methods and general laws, a new class of compounds—alcohols, triple substances which have no analogs in mineral chemistry, and despite it they have been formed exclusively by the play of affinities (...). Thus, the barrier between organic and mineral chemistry has now finally fallen down.

According to Berthelot organic chemistry is governed by the same laws as inorganic chemistry; this judgement might be considered as the final rejection of the concept of vital force. None the less, Berthelot outlines some limits of chemistry; at the end of his book he writes.

5.7. CHEMICAL EQUILIBRIUM AND REACTION RATE

The chemist will never intend to shape in his laboratory a leaf, fruit, muscle or vessel. These problems refer to physiology. This science discussed them, reveals the laws of the development of organs or—speaking more precisely—the laws of the development of the complete living organisms without which no separate organ could exist or have a proper environment for growth. But chemists (...) can undertake to produce substances contained in the living organisms [191].

We can admit, however, that the belief in the unity of the mineral and animate worlds, which was so deeply rooted in the minds of the alchemists and then undermined by the concept of vital force, began to dominate in science once again in the second half of the 19th century.

5.7. Chemical Equilibrium and Reaction Rate

From the mid-19th century, chemists paid increasing attention to the role of time in the course of chemical reactions. We have already seen that in ancient and mediaeval times a rapid change meant an easy one, i.e. the more readily one substance is transformed into another one, the quicker it occurs. However, those observations were related to physical changes and changes of inorganic compounds. The observations of syntheses and reactions between organic compounds made it possible to gain a better insight into the role of time. Berzelius made one of the first important observations in this respect. In 1836, he discovered a new force which—apart from the still enigmatic affinity and the more concrete influence of the mass of substances— changed the reaction rate. Investigating the process of obtaining ether from ethyl alcohol and oxidation of that alcohol to acetic acid, Berzelius noted that the former reaction ran faster in the presence of sulphuric acid, whereas the latter ran faster in the presence of platinum. Also, the decomposition of hydrogen peroxide into water and oxygen occurred more rapidly in the presence of platinum. He commented on the phenomenon in the following way:

It is, then, proved that several simple or compound bodies, soluble and insoluble, have the property of exercising on other bodies an action very different from chemical affinity. By means of this action they produce in these bodies decomposition of their elements (rather: decomposition into their elements— R.M.) and different recombinations of the same elements, to which they remain strangers.

This new force, which was unknown until now, is common to organic and inorganic nature. I do not believe that it is a force entirely independent of the electrochemical affinities of matter; I believe, on the contrary, that it is only a new manifestation, but since we cannot see their connection and mutual dependence, it will be easier to designate it by a separate name. I will call this force *catalytic force*, similarly I will call the decomposition of bodies by this force *catalysis*, as one designates the decomposition of bodies by chemical affinity *analysis* [192].

Later, it was observed that the reactions of organic substances do not always run to completion. In the year 1851, on the basis of the investigation of the etherification of alcohols in the presence of diluted sulphuric acid, A. W. Williamson wrote:

The transition from the static point of view to the dynamic consists in adding the measurements of time to the measurements of space. There is much evidence that time is necessary for chemical action, but this generally recognized fact is not taken into consideration when phenomena are explained [193].

Williamson's statement was not entirely correct. A year earlier, Ludwig Wilhelmy, while investigating the decomposition of saccharose into two monosaccharides in the presence of an acid, had found that the decomposition rate of sugar was directly proportional to the sugar content and to the acid content, the latter being a catalytic agent. On this basis he derived a mathematical formula which shows that the amount of substrate—saccharose—decreases exponentially with time.

The rate of substrate decay depends, according to Wilhelmy, on the 'factor of sugar conversion', a quantity similar but not identical, to the currently used reaction-velocity constant [194]. Wilhelmy's work was unnoticed not only by Williamson, who in fact carried out his investigations simultaneously with the German scientist, but by later chemists as well.

In the following years (1852–1854), during a study of etherification, Williamson also noticed that a mixture of substrates and products was always obtained as a final result, regardless of whether alcohol and sulphuric acid or ether (alkyl sulphuric acid) and water were mixed. He thus came to the conclusion that a definite equilibrium

5.7. CHEMICAL EQUILIBRIUM AND REACTION RATE 231

state between substrates and products is obtained due to the reaction. However, he did not consider this equilibrium as a state in which nothing happens; he held that the reaction runs simultaneously in two directions [195]. That observation laid the basis for a science of chemical equilibrium. Thus in the mid-19th century the problem of reaction kinetics was closely linked with the problem of chemical equilibrium.

In the years 1862-1866, M. Berthelot and L. Péan de Saint-Gilles carried out extensive measurements of the rate of the process of esterification of alcohols with organic acids. They noticed that:

the affinity of alcohols and esters is already manifested at room temperature, but generally at a slow rate. Each rise in temperature causes acceleration of the reaction [196].

On the basis of this measurements, Berthelot also proved that the quantity dy of ester being formed at a given moment depends on the ratio of ester existing at a given moment y to the maximum quantity of ester l which can be obtained under define conditions. Deriving a mathematical equation $dy = k(1-y/l)^2$, the investigator stated the agreement of the calculated results with the experimental data [197]. In the above equation k is a coefficient that is dependent on reaction, identical with the presently used reaction-velocity rate. Thus Berthelot justified Berthollet's observation published in 1803, which we have discussed on page 224.

Several years later, the Polish chemist Józef Jerzy Boguski derived an analogous equation which described the rate of dissolution of marble in diluted hydrochloric acid [198].

In the third part of their work dealing with the esterification of alcohols, B. Berthelot and L. Péan de Saint-Gilles explicitly formulate concepts concerning the role of velocity and equilibrium in chemical reactions:

Every time when acid and alcohol come into mutual contact, their compound is formed more or less readily, depending on the physical conditions of the experiment. Acid and alcohol are gradually neutralized and give room to new products: water and ester. However, when the proportions of the latter increase at the cost of the initial substances, we notice that action is slower and slower approaching continuously the stability state, where it is terminated and shows no differences observable in the experiment.

This state corresponds to the complete saturation of acid by alcohol. When it is accomplished, a mixture of the following four bodies is formed: alcohol and acid—initial compounds, and ester and water—newly formed bodies; the relative proportions of these four substances remain from that moment on unchangeable, unless the reaction conditions are changed. The same phenomena, but in inverse direction are observed when ester is decomposed by water [199].

The above considerations on the kinetics of chemical reactions and the equilibrium state were next continued by the Norwegian investigators: Cato Maximilian Guldberg and Peter Waage, who gave them a mathematical form. Since they made an important contribution to the development of chemistry and authors of chemistry coursebooks refer to them quite frequently but rather imprecisely, we shall now trace in some detail the views of the Norwegian chemists. They published two papers on this subject: a communication to the Norwegian Academy of Sciences published in French in 1867, and an article published in the *Journal für praktische Chemie* in 1879. The first publication is concerned with both the reaction of synthesis $A+B = AB$ and the reaction of simple exchange $AB+C = A+BC$. According to the authors both reactions occur under the influence of the resultant force of all attractions between the particular constituents of a mixture:

At a definite temperature this force can be considered as constant; its quantity will be denoted by k, which will be called a coefficient of affinity for this reaction. In both chemistry and mechanics the most real method will be determination of forces in their equilibrium. It means that we must study these chemical reactions in which forces that form new compounds are equalized by other forces. This happens in incomplete chemical reactions—in partial reactions, i.e. in reactions where (a) bonding and decomposition can occur simultaneously, and (b) exchange and reconstitution can occur simultaneously [200].

Next, denoting the active masses of substrates by symbols $p, q, ...,$ and the active masses of products by $p', q', ...,$ Guldberg and Waage point out that the forces with which substrates interact, equal to $k \cdot p \cdot q$ ..., must be equal to the forces with which the products interact—$k' \cdot p' \cdot q'$ ...

5.7. CHEMICAL EQUILIBRIUM AND REACTION RATE

It follows from the definition provided by the authors that those 'active masses' represent the numbers of atoms of a given substance in a unit volume, i.e. correspond to the volume molar concentration.

Guldberg and Waage go on to say in their paper that if the coefficient of affinity equals zero for any pair of substances, the substances do not interact. Neither will reaction occur when the molecules of a substance remain outside the area of interaction due to excessive dilution. On the other hand, forces with which the two substances react upon each other depend, according to the authors, on the volume molar concentrations (molarity) and the 'coefficient of affinity'. The sum of the products of those values for all the pairs of substances present in the solution, including the solvent—or more precisely—diluent, gives a resultant force T. The reaction rate v is proportional to that resultant force $T : v = \varphi T$, where φ denotes the coefficient of velocity.

The German article of Guldberg and Waage from 1879 was to a great extent a summary of the French publication and it was aimed at making its content accessible to a wider group of chemists. We can find there, however, a significant statement:

A chemical force with which two substances A and B act upon each other is equal to the product of their active masses multiplied by the activity coefficient (...). By the activity coefficient is understood a coefficient which depends on the chemical nature of both substances and temperature [201].

In this work, the authors define an 'active mass' as the mass of the substance in the activity area, and as the "amount of a given substance occurring in a unit volume". As we have seen in the quoted fragment of the publication from 1867, this concept may be identified with molarity. The conclusions of Guldberg and Waage discussed above formed the basis of a law referred to as the 'law of mass action'. The authors referred this name to the equation describing the dependence of the reaction rate on active masses. In his doctoral dissertation written in Wilhelm Ostwald's laboratory in 1898, Mieczysław Centnerszwer also refers to kinetic equations as to the law of mass action.

However, let us return to the final conclusions of Guldberg and Waage:

Equilibrium which accompanies that kind of chemical processes (i.e. the transmutation of substances A and B to substances A′ and B′—R.M.) is a state of mobile equilibrium because two inverse reactions occur at the same time since not only two new substances A′ and B′ are formed, but at the same time substances A and B are reproduced. When equal amounts of both pairs are formed in a unit of time, equilibrium is established [201].

From the rates of a given reaction and an opposite reaction occurring in the equilibrium state follows an equation of an equilibrium constant. This equation is often but not quite precisely referred to as the law of mass action.

Guldberg and Waage thus considered the equilibrium state as a dynamic state which only statistically brings about symptoms of concentration stability. Let us note, too, that the concept of velocity which since ancient times had been associated with the system's ability to transmute, finally led to the concept of chemical equilibrium, i.e. a state in which the resultant reaction rate equals zero, and the system as a whole is not able to undergo further transmutations.

It becomes quite evident from the cited statements that in the 1870s various investigators were aware of the fact that reaction rate and equilibrium state depend on temperature and concentration, i.e. in the case of gases it depends on pressure. This dependence was concisely formulated in 1884 by the French scientist Henry Louis le Châtelier as the so-called 'principle of mobile equilibrium' or le Châtelier principle. He extended to reactions which are in the equilibrium state van't Hoff's statement that an increase in temperature brings about disintegration of the products of exothermic reactions and their formation in the case of endothermic reactions [155, p. 161]. In the same year le Châtelier wrote:

Each system of a stable chemical equilibrium subjected to the effect of an external stimulus, which tends to change either its temperature or condensation (pressure, concentration, number of molecules in a unit of volume), in its whole or only in any part, is submitted only to such internal transformations, which—if they occurred separately—would cause change in temperature or condensation with a reverse sense than that of a change caused by an external stimulus [202].

In his work of 1884 van't Hoff expressed the kinetic equations of reactions in a form which has been used until the present day.

Following his practice, the constant k, which was used in his equations, has been called a 'reaction-velocity constant'. Simultaneously, he introduced the concept of molecularity of reaction: the number of molecules taking part in it. This concept became a basis for the investigation of reaction mechanisms. If the molecularity of a given reaction determined on the basis of measurements of reaction rates is discordant with that predicted on the basis of a stoichiometric equation, we may be sure that the reaction does not proceed directly but it proceeds through indirect stages. Van't Hoff also stated that this kind of conformity can be observed only for a relatively small number of reactions, i.e. there are almost always factors which disturb the course of reaction. These factors included, according to van't Hoff, the effect of solvents, and temperature effects. Thus, the value of equilibrium constant K resulting from the law of mass action depends on temperature. If we change the temperature of a system in the equilibrium state, the reaction will proceed in a direction which permits it to achieve a new equilibrium state. Van't Hoff proved that a temperature change of the logarithm value of the equilibrium constant is directly proportional to the heat Q given off during a change and to the difference of the inverse of the temperatures:

$$\ln K_{T_1} - \ln K_{T_2} = (Q/R)(T_1^{-1} - T_2^{-1}).$$

Van't Hoff's considerations once again link the course of chemical reaction with the heat liberated. Such a hypothesis had already been propounded by Lavoisier and Laplace, but their suggestion that this heat was associated with the affinity forces failed to draw a response from the chemists in the early 19th century. The latter's attention was directed mainly to the synthesis and structure of chemical compounds.

5.8. Is it Really the Heat of Reaction?

A suggestion analogous to that of Lavoisier and Laplace was proposed in the years 1839/40 by Germain Henri Hess. He carried out measurements of the heat emitted during the dilution of the aqueous solutions of sulphuric acid of different concentrations by stoichiometric aliquots of pure water. Treating the solutions of sulphur trioxide in the gradually increasing volumes of water as a series of compounds, in accordance with the law of multiple proportions, Hess stated: "when a compound is formed, the amount of emitted heat is constant regardless of the fact whether the compound is formed directly or indirectly with different restarts (i.e. successive dilutions)" [203].

This law holds only under isobaric conditions and when, except for chemical work, no other work is done.

However, in the year 1854 the Danish chemist Julius Thomsen explicitly set out the idea of reaction heat as a measure of affinity. He stated:

(...) it appears that the affinity of two bodies is manifested by an ability of direct combination; upon association an amount of heat corresponding to affinity is given off [204].

Thomsen's statement, like the statement of Lavoisier and Laplace can be applied only to changes which are accompanied by the liberation of heat. However, the cause of reactions that absorbed heat was still unknown.

A similar view, it seems, was shared by Butlerov. The Russian chemist, as we have stated in the preceding chapter, was mainly concerned with the structure of chemical compounds, and he called 'units of affinity' what we now refer to as units of valency. Therefore, he gave a more extensive definition of the term 'affinity' than that provided by Thomsen or Berthelot. At the 36th Congress of German Scientists and Physicians in 1861, Butlerov said:

The amount, of affinity must be distinguished from its force—a greater or smaller energy with which it combines bodies with one another. This force varies depending on the nature of reacting bodies and on the conditions under which a reaction takes place [147].

Next, he suggested that if four hydrogen atoms were added one after the other to a carbon atom, the ease of addition of each of them would depend on the number of hydrogen atoms previously added.

Further advancement in the understanding of the role of energy in chemical transmutations is due to Berthelot. On the basis of his earlier publications from the years 1862–1866, he came to the conclusion that when the amount of heat liberated during a chemical

5.8. IS IT REALLY THE HEAT OF REACTION?

change is to be determined, the system must not be supplied with external energy, i.e. the system must be isolated. He realized that part of the heat measured by thermochemical methods might come from mechanical or electrical work performed on the system. He formulated this idea for the first time in 1867, and then he gave a comprehensive view of his conception in a textbook entitled *Essai de mechanique chimique fondée sur la thermochimie* published in Paris in 1879. He presented his conception in the form of three principles:

I. The principle of molecular work—the amount heat given off during any reaction is a measure of the sum of chemical and physical work done in the course of this reaction. This principle provides a measure of chemical affinity.

II. The principle of thermal equivalence of chemical transformations, or in other words the principle of initial and final state—if physical or chemical changes occur in a system of simple or compound bodies under definite conditions and are capable of reaching a new state, so that no external mechanical phenomenon occurs in the system, then the amount of liberated or absorbed heat due to change depends exclusively on the initial and final state of the system; it remains the same irrespective of the character and sequence of intermediate states.

III. The principle of maximum work—each chemical change accomplished without the action of external energy tends to produce a body or a set of bodies with most heat given off [205].

These three principles are supplemented in the textbook with the following statements:

Each chemical reaction which can be effected without initial work and without the action of external energy on the bodies present in the system must occur (spontaneously—R.M.) if it gives off heat.

Heat absorbed in the process of decomposition of a body is precisely equal to that given off in the course of formation of this same compound provided that the initial and final states are identical.

Thus, Berthelot significantly modifies the attemps of Lavoisier and Laplace, as well as those of Thomsen, which were aimed at associating reactivity with liberated heat. He emphasizes the role of a factor which today may be referred to as chemical work, and besides takes into account processes that occur during heat absorption; however, for him heat is still a measure of affinity.

In all the formulations from the latter half of the 19th century,

5. CAPACITY OF A SUBSTANCE

which were aimed at explaining the cause of chemical processes and developing a quantitative approach to them, the influence of advancement in physics, and particularly in one of its branches—thermodynamics, is obvious. As we have mentioned in Chapter 2, in 1847 Helmholtz formulated the 'principle of conservation of force', or the first principle of thermodynamics which shows the equivalence of different forms of energy. These forms then included mechanical, electrical, and thermal energy. In fact, this principle resulted from N. L. S. Carnot's considerations on the heat engine from 1824.

However, it was only in 1850 that the significance of those considerations was shown by Clausius who continued Carnot's work, and in 1865 introduced the concept of 'entropy'. In the following year, Boltzmann provided a statistical interpretation of this concept. Thus Berthelot's considerations quoted above permitted him to consider chemical energy (the energy of chemical transmutations) as another form of energy.

The next step in this field was made, independently of each other, by Willard Gibbs in 1876 and Hermann Helmholtz in 1882. Both scientists stated that not all the amount of energy contained in an investigated system can be transformed into work (today we formulate it more precisely: it cannot be given up to the environment in the form of work). Part of this energy, connected with the change of entropy occurring during a transmutation, must be released at a given temperature to the environment in the form of heat. Therefore, the maximum work which can be achieved during a transmutation must be smaller than the change of the total enthalpy of the system ΔH or the whole internal energy of the system ΔU by the magnitude of the product of entropy change ΔS and absolute temperature T at which a transmutation occurs.

In 1886, Pierre Duhem introduced the concept of 'thermodynamical potentials'; they vary depending on the conditions under which the process is developed, i.e. under the condition of constant temperature or constant enthropy, and constant volume or constant pressure. Thermodynamic potentials refer to the basis thermodynamic functions, i.e. internal energy U, enthalpy H, the Gibbs function $G = H - TS$, called "free enthalpy', and the Helmholtz function $F = U - TS$, called 'free energy'. The adjective 'free' is used to show that it is that part of enthalpy or the internal energy of the system which can be, at a given temperature, exchanged with the environment in the form of work.

It is the changes in the values of the two thermodynamic potentials F and G that decide whether a reaction will proceed spontaneously and not the heat liberated, as was suggested by Lavoisier and Laplace, and later by Thomsen and Berthelot. It was then realized that a system is in equilibrium not when its internal energy or enthalpy has the smallest value and entropy has the greatest value, but when the free enthalpy (isobaric transformations) or the free energy (transformations at constant volume) has the smallest value. Decrease in potentials F and G can be achieved by both a decrease in the value of enthalpy (or internal energy) and a increase in entropy occurring during the transformations. The prevalence of one of the factors may suffice to initiate a spontaneous process. Thus it is possible to bring about a transformation in which entropy will be decreased if enthalpy (or internal energy) is also decreased to a sufficient degree. Similarly, it is possible to obtain a transmutation in which the value of enthalpy or internal energy will be increased, i.e. heat will be absorbed if a sufficiently high increase of entropy accompanies the process.

Most chemical processes occur under isobaric and near-isothermal conditions. Therefore, of all thermodynamic potentials free enthalpy G has the greatest importance. In 1884, van't Hoff considered changes in free enthalpy which occurred in the process of chemical transmutations. He came to the conclusion that, for a reaction to proceed spontaneously, the change in potential must be negative. He proposed calling chemical affinity the negative difference of the value of free enthalpy of the investigated system with given concentrations of the constituents and a minimal value of this potential for the system in which the same constituents are in equilibrium.

In this way van't Hoff defined precisely, quantitatively, the value of affinity; and besides he combined the two factors, considered from ancient times, which characterize a chemical change: affinity, which predetermines the ease of transmutation, and energy, obtained in the process of transmutation. The Dutch investigator showed at the same time that the equilibrium constant, which Guldberg and Waage had found through a kinetic approach, could also be derived from thermodynamic investigations. For that purpose he employed another thermodynamic magnitude, introduced by Gibbs a few years earlier, and called a 'chemical potential'. This is equal to the derivative of particular thermodynamic potentials with respect to the number of moles of a given compound which undergo a reaction in relation to the determined pairs of parameters, which vary for different thermodynamic potentials. Precisely, chemical potential does not play the role of potential but it is a generalized force [206], and like a mechanical force causes a body to move, chemical potential—a real 'chemical force'—causes change in the number of moles of a given substance, i.e. causes a chemical reaction. In the equilibrium state the sum of the products of the chemical potentials of each constituents and the corresponding stoichiometric coefficient in an equation of a chemical reaction must equal zero. This condition can be fulfilled, as A. F. Horstman showed, only when the expression containing products of the molar concentrations of the mixture constituents has a form identical with that in the law of mass action formulated by Guldberg and Waage [207].

As a matter of fact, in spite of associating affinity with enthalpy it is still an open question which quantity is a measure of the stability of a compounds: is it liberated heat which is equal to the change in the system's enthalpy under isobaric conditions, or is it affinity which shows in what degree the system is remote from the equilibrium state? In this case the so-called 'standard affinity' is taken into consideration; the equilibrium state is compared with a state in which molar concentrations of all reagents are equal to unity. Let us note, however, that the affinity associated with the equilibrium constant points to the 'resistance' of the equilibrium state to changes in the concentration of constituents, but only under isothermic conditions. The higher the value of the free enthalpy of change in the equilibrium state, i.e. the higher the value of the equilibrium constant, the greater the amount of the product that will disintegrate in isothermal—isobaric conditions with decreasing substrate concentration. However, it follows from van't Hoff's isobar that the greater the value of the enthalpy of reaction

$\Delta H = Q$, i.e. the greater the difference between the total enthalpy of and the total enthalpy of substrates products, the greater the quantity of the product that will disintegrate when the temperature increases by a definite value under isobaric-isentropic conditions. As in ancient times, we must consider two magnitudes which determine the system's ability to react, i.e. we must decide upon the stability of the products. However, these magnitudes refer to different conditions under which transmutations occur.

5.9. Catalysis and Activation Energy

On page 230 we have quoted Berzelius' statement from 1836 in which he introduced the term 'calatysis'. Since that time catalytic processes have been regarded as very common phenomena; it has been found that almost every additional constituent affects the reaction rate. In 1839, Liebig recognized that the reaction rate is facilitated by the state of aggregation, the 'element' of water as well as "a contact with a third body which does not enter the compound" [208]. In 1901, Wilhelm Ostwald provided the following definition of a catalyst:

Fig. 19. Wilhelm Ostwald,
acc. to W. Strube, *Der historische Weg der Chemie*, Leipzig 1976.

A catalyst is a body that changes the rate of a chemical reaction but is absent from the final products of that reaction [209].

The effect of catalysis was little understood for a number of years. Besides, there was a widespread misconception that catalysts do not take part in reactions. However, the significance of this statement corresponded to Liebig's formulation in a sense that no traces of a catalyst are found in the composition of the final products. It was not until the 20th century that it was realized that a catalyst does take part in a reaction, but eventually it is restored—theoretically with no losses.

In 1889, Svante Arrhenius carried out an investigation into the desintegration of saccharose at different temperatures and found a nonlinear increase in the decomposition rate constant, dependent on temperature. With the help of the known statistical Boltzman law which describes relative numbers of systems with different energy dependent upon temperature, he presented the dependence of a reaction-velocity constant upon absolute temperature T in the form $K = A\exp(-E/RT)$. Quantity E was interpreted as an activation energy, which is a kind of barrier hampering the course of a chemical reaction. It is not until the system reaches a potential enabling it to surpass this barrier that thermodynamic differences between substrates and reaction products begin to gain importance. On this basis it was realized that the action of a catalyst consists in decreasing the activation energy. It was found later that some substances—inhibitors—hamper the course of reaction by increasing the activation energy.

Sometimes reaction products turn out to be catalysts that speed up its rate. Such reactions are referred to as autocatalytic; their kinetics were explained in 1916 by the Polish physical chemist Jan Zawidzki [210].

In 1917, Walter Nernst proposed for the first time the mechanism of a chain reaction in order to explain the process of obtaining hydrogen chloride from chlorine and hydrogen. In the 1920s Cyril Norman Hinshelwood in England and Nicolai N. Semenov in the USSR showed that a considerable number of known reactions undergo intermediate stages, and that in reality they are chain reactions. The work of these scientists permitted discovery of the kinetics of such reactions. All reactions with the participation of catalysts are chain reactions. At present we have also realized that enzymatic reactions, which play a significant role in biology, are also chain reactions because enzymes are biological catalysts.

5.10. Irreversible Thermodynamic Reactions

A substantial portion of the considerations on thermodynamics at the turn of the last century and in the early 20th century, including considerations on chemical processes, concerned 'reversible thermodynamical processes', i.e. infinitely slow, idealized processes occuring when a system remains constantly in equilibrium states. However, such processes are not encountered in nature. All processes which occur in nature are thermodynamically irreversible, i.e. their direction can be reversed only by a change in the conditions under which a given phenomenon occurs. In order to reverse the course of a chemical transmutation, it is necessary to change the temperature of a system or raise the concentration of one of its constituents. It was not until the mid-19th century that Rudolf Clausius and William Thomson (Lord Kelvin), while investigating the effect of the heat engine, found that the efficiency of every real engine is smaller than the efficiency of an engine in which reversible thermodynamic processes would occur. It followed from this observation that in the case of an ideal engine the total value of the entropy of a gas in the engine and of the environment would remain unchanged, whereas in the case of a real engine this value must increase during the engine's operation. Nonetheless, this statement was not sufficient to fully develop the thermodynamics of real, thermodynamically irreversible processes.

Władysław Natanson, a professor at the Jagiellonian University in Cracow is considered to be one of the founders of the modern thermodynamics of irreversible processes. In the years 1896–1897 he published several papers on this subject.

Irreversible thermodynamic processes occur in systems which remain in states different from equilibrium; therefore there is a certain heterogeneity in such systems which acts as a thermodynamic stimulus.

If the state of a system is not too remote from the equilibrium state, such a stimulus will cause certain quantities to flow (e.g. mass or energy), their flowrate, i.e. the flow value per unit time, being directly proportional to the magnitude of the stimulus. However, if the state of a system is very remote from the equilibrium state, the value of the flow rate is a nonlinear function of the stimulus value: it thus depends on the value of a stimulus not only in the first power but in higher powers as well. In such a case sources of entropy may appear in the system.

In the 1950s, Ilya Prigogine applied this theory to biological processes; biological systems are very far from the equilibrium state, therefore they are nonlinear thermodynamically irreversible processes. Thus the equations that describe such processes must include more elements than equations that describe reversible processes; they must include additional elements which describe flows. We have pointed out earlier that the majority of chemical processes was regarded—with sufficient agreement of theory and practice—as reversible thermodynamic processes. Biological processes must be considered as thermodynamically irreversible processes. It follows that chemical processes can be described by means of simpler laws and formulae than biological processes.

If we analyse the historical changes of relations of chemical processes to biological processes, we notice cyclical changes of opinions. It was believed in the Middle Ages, as we have seen, that both kinds of processes were subject to the same rules and they differed only in the time scale. In the 17th–19th centuries, differences between these processes were accounted for by vital force. In the mid-19th century, it was recognized once again that the same laws govern processes of both kinds, and the present-day thermodynamics of irreversible processes attempts to explicate the difference in the behaviour of most frequently observed chemical and biological systems. Let us note that Prigogine showed that the output of entropy sources in living organisms decreases with ageing, hence the irreversibility of biological processes decreases too, and they thus approximate reversible processes. Therefore, with the development of an organism, processes referred to as biological are gradually replaced by processes commonly known as

5.10. IRREVERSIBLE THERMODYNAMIC REACTIONS

chemical. As far back as 1804, Jędrzej Śniadecki wrote in his 'Theory of Organic Entities':

> (...) we have shown that all organic processes depend upon the effect of organizing forces and their superiority over counter-organic forces in the same way as the superiority of desorganizing forces depends upon the prevalence of forces of affinity; thus it appears that in the first half of life organic forces are still predominant and control life whereas in the latter half they gradually grow weaker, and thereby they give more and more way to the forces of affinity until, eventually, they surrender themselves completely to such forces, and thereby all organic processes and life are terminated [190, p. 185].

Thus, we can see that the old views about the existence of organizing forces contained the seeds of ideas whose significance we are only now beginning to realize.

Now we seem to have understood the relations between an ability to react and a reaction rate, and the relations between this ability and the structure of reacting molecules. As we have mentioned it was believed for many centuries that an 'easy' reaction meant a 'quick' reaction. Today we know that an ability to react depends on the difference between the free enthalpy of reaction products and substrates, whereas its rate depends on activation energy, which occurs in an equation proposed by Arrhenius (cf. p. 242). The quantitative composition of a mixture corresponding to chemical equilibrium thus depends on factors other than the velocity with which this state is reached.

The structure of a substance was for many centuries associated with an ability to react. As late as the mid-19th century the term 'affinity' was applied to both notions. It was only thanks to the development of thermodynamics that the denotation of this term was precisely referred only to an ability to react. However, as we have emphasized in the conclusion of the preceding chapter, the methods of quantum mechanics, which permit us to calculate the distribution of electron density in molecules, even in the state of their interaction, give some indications about both structure and ability to react. Thus the two problems of chemistry have again closely overlapped.

Postscript: a Reflection

In the foregoing five chapters we have traced the development of chemical thought, which is an element of the development of human thought. Each idea, each outlook from ancient to modern times has carried an element of scientific truth and has left a trace in our culture. No wonder: these ideas and views were to present, make concrete and partly explain the observed complex phenomena occurring in nature—in the sky and on the Earth, among the stars, in the kingdom of minerals, plants, animals, and in man himself.

While new information was acquired these views underwent modifications; they frequently ceased to conform with the prevailing beliefs and then they were rejected. They were sometimes rejected, however, together with the positive elements they contained. After years, or even centuries, these views were recalled: they were brought back in a new, more appropriate form. Such was the fate of Aristotle's search for the elementary properties, the holistic approach to nature assumed by the ancient philosophers and alchemists, the vital force theory, Berzelius' dualistic theory, and many others.

Science is a process of continuous and gradual development; its progress being neither uniform in time nor in its various branches. Sometimes, in some branch of science, progress is made very rapidly. It seems to us from the perspective of time that old views were renounced and all of a sudden new ones were adopted. Some historians and philosophers of science call such a situation a scientific revolution. However, upon closer inspection we shall notice that 'new' views have been gradually formed; and they still are.

Science has never come to a dead stop. The reader will have sufficiently realized by now that everything that is being taught in schools and at universities is merely the 'present state of our knowledge'.

Chronological Tables

Table I. The Period of Practical Arts

Years	Period	Tools and products	Practical arts and methods of processing	Cultural and technological achievements
1	2	3	4	5
until c. 7000 BC	Late Palaeolithic Period (Older Stone Age)	Stones and tools made of chipped stone, bows harpoons	Fire cooking Beginnings of tanning Nets and baskets	Speech myths Naturalistic painting and sculpture
7000–2000 BC	Neolithic Period (Younger Stone Age)	Tools made of polished stone Ornaments from raw gold and copper	Agriculture Pottery Weaving baking and boiling Crude Carpentry	Primitive therapeutics and surgery Calendar for farmers Myths about creation
2000–1200 BC	Bronze Age	Canals and dams Sailing boats Wheeled carts	Casting of copper and bronze Riveting and welding Polished ceramics	Numbers and writing Scales and measurements Arithmetic and geometry Formation of first states: Egypt, Babylonia, Persia, and Judea 1800 — Hammurabi's Codex
1200– c. 600 BC	Early Iron Age	Iron Glass Plough Water wheel pumps	Preparation of drugs and dyes	Alphabet Metal money c. 800 — Homer's epopees 750 — foundation of Rome

* The chronological tables have been compiled from the tables in:
Bernal J., 'Science in History', Watts and Co., London 1954.
Mierzecka A., *I uczeni są ludźmi* (And Scientists Are Humans), Wiedza Powszechna, Warszawa 1962.
Valentin H., *Geschichte der Pharmazie und Chemie in Form von Zeittafeln*, Stuttgart 1950.
Neufeldt S., *Chronologie Chemie 1800-1970*, Verlag Chemie, Weinstein 1977.

Table II. The Period of Greek Science

Years	Political and social developments	Cultural developments	Scholars	Exact and natural sciences	Chemistry and the structure of matter
1	2	3	4	5	6
600–300 BC	Greek republics		Thales of Miletus (620–540)	Formation of philosophical schools (Ionian and Eleatic Schools)	Water as primary matter
			Anaximenes (585–525)		Air as primary matter
			Pythagoras (572–497)	Harmony of the world	
			Heraclitus (540–480)	Idea of permanent change	
			Empedocles (483–423)		Four primary matters
			Hippocrates (460–377)	Beginnings of scientific medicine	
			Democritus (460–370)		Atomism
		387—Plato's Academy	Plato (427–370)	Objective cognitive idealism	
	336–323 Alexander the Great's empire	335— Aristotle's school	Aristotle (384–322)		Four elements
300–200 BC		290— museum in Alexandria 279— ligthouse at Pharos	Archimedes (287–212)	260— description of operation of simple machines	

Table II continued

1	2	3	4	5	6
200–0 BC	Rome's domination over the Greek world		Bolos of Mendes Lucretius (95–55) Celsus (53–7)	Development of medicine	Beginnings of alchemy *De rerum natura*
0–500	Roman Empire Migrations of peoples 476—fall of Rome		Ptolemy (100–168) Galen (130–200) Zosimos (350–420)	150—geocentric theory further development of medicine	28 volumes of alchemical works Alexandrian school of alchemy
500–700		529—fall of Plato's Academy Fall of Alexandria Development of Arabic culture in the East			
700–900	Charlemagne (768–814)	Development of Arabic science in Spain	Geber (720–813)	773—Arab numbers	Non-organic acids Alchemic elements
900–1100	966—beginning of the Polish state		Avicenna (980–1037)		
1100–1200	Crusades	Paper in Spain Translation from Arabic into Latin			

Table II continued

1	2	3	4	5	6
		Universities in Bologna, Paris, and Oxford			
1200–1300		Tower clocks Gunpowder	Albertus Magnus (1193–1280) Roger Bacon (1214–1294)		Sulphur and mercury— philosophical constituents of metals
1300–1400		Spectacles Compass 1364— foundation of Cracow University	R. Lullus (1235–1315) Arnold de Villanova (1235–1311) Ockham (Occam) (1300–1349)		
1400–1500	1492— discovery of America	Development of metallurgy Platonian Academy in Florence 1450— Gutenberg's printing press	Leonardo da Vinci (1452–1519)		
1500–1600	Reformation	Lunettes	Copernicus (1473–1543) Paracelsus (1493–1541) Agricola (1494–1555) Ercker (1530–1594) Gilbert (1540–1603) Libavius (1540–1615)	Heliocentric system Earth as a magnet	Beginnings of iatrochemistry *De re metalica* 1574— *Compendium* *Alchemia*

Table III. The Period of Quantitative Research

Years	Political and cultural developments	Scholars	Advancement in Technology Exact and Natural Sciences	Chemistry and the structure of matter
1	2	3	4	5
1600–1650	1603—Academia dei Lincei Thirty-Years War	Gallileo (1564–1642) Kepler (1571–1630) Sennert (1572–1637) Van Helmont (1577–1644) Gassendi (1592–1655) Glauber (1604–1668)	1610—Discovery of the Jupiter moons 1600—laws of circulation of planets Vital force theory	Rebirth of the corpuscalar view of the texture of matter Discovery of carbon dioxide Glauber's salt
1650–1700	1662—Royal Society 1669—Academie des sciences 1683— —Poland's Vienna Victory over the Turks	Boyle (1627–1691) Huygens (1629–1695) Becher (1635–1682) Newton (1643–1727)	 Pendulum clock Wave theory of light Differential calculus 1682—law of universal gravitation	1661— *The Sceptical Chymist* Investigations of compressibility of gases Sulphur as combustible earth

Table III continued

1	2	3	4	5
1700– –1750	Manu- factories in France 1749—Scien- tific Society in Gdańsk	Lemery (1645–1715) Stahl (1660–1734) Boerhaave (1668–1738) Geoffroy (1672–1731) Lomonosov (1711–1765)	1707—Newton's 'Opticks' 1714—Fahrenheit's thermometer 1710—porcelain 1728—Bernoulli's theory of gases	*Cours de chimie* 1697–1717— phlogiston theory 1718—Geoffrey's table of affinities 1746—chamber method of making sulphuric acid
1750– –1800	1751–1772 Great French Encyclo- paedies Beginnings of the factory system Formation of the USA Partitions of Poland French Revolution	Priestley (1733–1804) Lavoisier (1743–1794) Berthollet (1748–1822) Rumford (1753–1814)	1760—calorimeter 1783—caloric hypothesis 1784—Coulomb's law 1796—Volta's battery 1798—kinetic theory of heat	1755—rediscovery of CO_2 1766—discovery of hydrogen 1772—discovery of nitrogen 1774—discovery of oxygen 1775—selective attraction 1777—combustion theory 1791—Leblanc's method of obtaining soda 1792—stoichiometry 1799—law of constant composition

Table IV. The 19th Century

Years	Scientific organizations	Scientists	Technological advances	Exact and natural sciences	General (physical) and mineral chemistry	Organic chemistry
1	2	3	4	5	6	7
1800–1850	1800—Society of the Friends of Science in Warsaw (1808)–1816—Warsaw University	Volta (1745–1827) Dalton (1766–1844) Śniadecki (1768–1838) Davy (1778–1829) Berzelius (1770—1848) Faraday (1791–1867) Dumas (1800–1884) Wöhler (1800–1882) Liebig (1803–1873) Helmholtz (1821–1894)	1814—steam engine 1839—beginning of photography	1800—infrared radiation 1801—ultraviolet radiation 1807—term 'energy' (Young) law of thermal expansion of gases 1808—polarization of light 1820—electromagnetism (Ampère, Oersted)	1800—'Principles of Chemistry' (Śniadecki) 1802—law of partial pressures (Dalton) 1808—'New System of Chemical Philosophy' (Dalton) Electrolytic production of sodium and potassium (Davy) 1811—Avogadro's theory	1804—'Theory of Organic Entities' (Śniadecki) 1820—isolation of alkaloids 1825—discovery of benzene (Faraday) 1826—making of aniline 1828—making of urea from cyanamide (Wöhler) 1831—elementary analysis (Liebig) 1832—concept of isomerism (Berzelius)

Table IV continued

1	2	3	4	5	6	7
				1824—Carnot's circular cycle	1815—hypothesis of complex structure of atom (Prout)	Theory of radicals (Dumas, Liebig)
		Clausius (1822–1888)				
		Cannizzaro (1826–1910)		1832—electromagnetic induction (Faraday)	1818—Berzelius' dualistic theory	
		Berthelot (1827–1907)		1833—laws of electrolysis (Faraday)	1836—concept of catalysis (Berzelius)	1834—derivatives of benzene (Mitscherlich)
		Butlerow (1828–1886)		1842—equivalence of heat and work (Meyer)	1838—hydrogen definition of acid (Liebig)	1840—fertilization of soil with mineral agents (Liebig)
		Kekulè (1829–1896)				
		Maxwell (1831–1879)		1843—Joule's heat	1840—Hess' law	1845—synthesis of acetic acid from elements (Kolbe)
		Mendeleev (1834–1907)		1847—the first law of thermodynamics (Helmholtz)		

1850–1900	1860—International Congress of Chemists in Karlsruhe 1873—Academy of Sciences in Cracow	van der Waals (1837–1923) van't Hoff (1852–1911) Ostwald (1853–1932) Skłodowska-Curie (1867–1934)	1856—Bessemer's converter 1853–1859 beginning of petroleum industry 1861—production of soda by the Solvay method 1864—vehicle propelled by an internal combustion engine 1867—electrical motor 1879—electric bulb	1850—the second law of thermodynamics (Clausius) 1857—kinetic theory of gases (Clausius, Maxwell) 1858—evolution theory (Darwin) 1859—spectral analysis (Kirchhoff, Bunsen) 1861–1865 electromagnetic theory of light (Maxwell) 1865—concept of entropy (Clausius)	1850—the role of time in the course of chemical reactions (Wilhelmy) 1852—valency (Frankland) Concept of chemical equivalence (Williamson) 1858—acceptance of Avogadro's theory (Cannizzaro) 1864—the law of octaves (Newlands) 1865—Loschmidt's number	1850—synthesis of amino acids 1853—unitary theory (Gerhardt) 1856—synthesis of artificial dyes (Natanson, Perkin) 1858—tetravalence of carbon (Kekulé) 1860—double bond (Erlenmeyer) 1861—structural formulae (Butlerov)

Table IV continued

1	2	3	4	5	6	7
			1895–1897 beginnings of radio broadcasting (Popov, Marconi)	Laws of heredity (Mendel) 1873—equation of real gas state (van der Waals) 1881—discovery of the electron (Stoney Helmholtz) 1888—electromagnetic waves (Hertz) Energetism (Ostwald) 1895—Röntgen's rays 1896—uranium radiation (Becquerel) 1898—electrons as a part of an atom (Thomson)	1867—law of mass action (Guldberg, Waage) 1869— Periodic system of the Elements (Mendeleev) 1887—theory of electrolyte dissociation (Arrhenius) Equation of chemical kinetics Conditions of spontaneous reactions (van't Hoff) 1889— thermodynamics of electrical cells (Nernst) 1898—discovery of polonium and radium (M. and P. Curie)	1865—structure of aromatic compounds (Kekulé) 1874— tetrahedral carbon (van't Hoff, Le Bel) 1888— investigations of porphyrins Development of biochemistry (Nencki) 1892—Geneva nomenclature of organic compounds

Table V. 20th Century

Scientists	Technological advances	Exact and natural sciences	General (physical) and inorganic chemistry	Organic chemistry
1	2	3	4	5
Planck (1858–1947)	1900—flight of a steered baloon	1900—quantum theory	1908—final acceptance of the conception of corpuscular texture of matter	1906—chromatography (Cwiet)
Nernst (1864–1941)	1903—beginnings of aviation	1905—photons equivalence of mass and energy; special theory of relativity (Einstein)	1912—theory of molecular dipoles (Debye)	1907—first plastic mass (bakelite)
Rutherford (1871–1937)	1919—prototype of mass spectrometer			1913—polyvinyl chloride fibres
Smoluchowski (1872–1917)	1921—attempts at telephotography	1910—determination of electron, charge (Millikan)	1913—mechanism of chain reaction (Bodenstein)	Petrol from coal
Einstein (1879–1955)	1924— ultracentrifuge (Svedberg)	1911—atomic nucleus	1916—theory of electron octet	1920—beginnings of macromolecular chemistry (Staudinger)
Hahn (1879–1968)	1930—beginnings of television	1913—Rutherford– –Bohr's model of atom	1920—hydrogen bond	1925—catalytic separation of saturated and non-saturated hydrocarbons (Fischer and Tropsch)
Debye (1884–1966)	1937—making of polyamide fibre		1923—generalization of the conception of acids and bases (Bronsted, Lewis)	1927—cryptions theory of orienting influence of substituents (Ingold)

Table V continued

1	2	3	4	5
Staudinger (1881–1965)		Law of displacements (Fajans, Soddy)		1929—proof of existence of organic radicals
Bohr (1885–1962)	first practical spectrophotometer for infrared	1919—artificial nuclear transmutation (Rutherford)	1925—polarography (Heyrovský)	
Heyrovský (1890–1967)	1939—teflon invented		1927—quantum chemistry (Heitler, London)	
Schrödinger (1887–1961)	1940—transuranium elements; Np and Pu	1924—de Broglie's waves		
de Broglie (1892–1987)	1942—atomic pile	1925—Pauli's exclusion principle	1928—Raman scattering 1929—discovery of isotopes	1933—conception of mesomery and chemical resonance
Heisenberg (1901–1976)	1946—computers	Quantum mechanics (Heisenberg, Schrödinger)	1930—theory of irreversible coupled thermodynamic processes	
Pauling 1901	1948—transistors	1932—discovery of positron and neutron	1932—scale of electronegativity (Pauling)	1935—mechanisms and kinetics of substitution reaction
Joliot-Curie F. (1900–1958)	1954—principle of masers and lasers	1934—artificial radiation (Joliot-Curie)		
Onsager (1902–1976)	1960—first man in outer space	1935—prediction of existence of mesons	1952—conception of donor-acceptor complexes CT (Milliken)	Crystal-like form of viurs tobacco mosaic
	1965—chemical laser 1969—man's landing on the Moon			1953—conception biochemical evolution;

1939—Bethe's cycle; formation of helium from hydrogen on the Sun

1945—electron paramagnetic resonance

1946—nuclear magnetic resonance

1955—discovery of antiproton

1964—conception of quarks

1961—isotope ^{12}C as the basis for determination of atomic masses

1962—compounds of noble gases

Formation of aminoacids in electrical arc

Discovery of DNA structure (double helix)

1966—discovery of genetic code

References

General References

[1] Asimov I., 'A Short History of Chemistry', Doubleday and Co., Garden City, New York 1965.
[2] Bernal J. P., 'Science in History', Watts and Co., London 1954.
[3] Bykov G. B., 'A History of Organic Chemistry' (in Russian), Nauka, Moskva 1978.
[4] 'Classical Scientific Papers—Chemistry', First Series, American Elsevier, London–New York 1968.
[5] 'Classical Scientific Papers—Chemistry', Second Series, Mills and Booms Ltd., London 1970.
[6] Dampier F., 'History of Science', University Press, Cambridge 1948.
[7] Ferchl F., Süssengut A., *Kurzgeschichte der Chemie*, Arthur Nemayer Verlag, Mittenwald 1936.
[8] Fierz-David H. E., *Entwicklungsgeschichte der Chemie*, Verlag Birkhäuser, Basel 1945.
[9] Giua M., *Storia della chimica*, Union Tipografico—Editrice Torinese, Torino 1962.
[10] Kuznetsov V. I., 'The Evolution of Concepts of Basic Laws of Chemistry' (in Russian), Nauka, Moscow 1967.
[11] Kwiatkowski E., 'A History of Chemistry and Chemical Industry' (in Polish), WNT, Warszawa 1962.
[12] Legowicz J., Ed., 'A History of Mediaeval Philosophy' (in Polish), PWN, Warszawa 1979.
[13] Losee J., 'A Historical Introduction to the Philosophy of Science', University Press, London—Oxford 1972.
[14] 'The Materialists of Ancient Greece' (in Russian), Dynnika M., Ed., GIPL, Moskva 1955.
[15] Ostwald W., *Leitlinien der Chemie*, Leipzig 1906.
[16] Partington J., 'History of Chemistry', MacMillan and Co., London 1970 (Vol. I), 1961 (Vol. II), 1962 (Vol. III), 1964 (Vol. IV).
[17] Russel C. A., in: 'Recent Development in History of Chemistry', The Royal Society of Chemistry, London 1985.
[18] Solovev J. J., 'The Evolution of the Fundamental Theoretical Problems of Chemistry', (in Russian), Nauka, Moskva 1971.

[19] 'Source Book in Chemistry 1400–1900', Eds. Leicester H. M., Klickstein H. S., McGraw-Hill Co., London 1952.
[20] Strube I., Stoltz R., Remane H., *Geschichte der Chemie*, VEB Deutscher Verlag der Wissenschaften, Berlin 1986.
[21] Trifonov D. N., Krivonomarov A. N., Lisnevskii J. I., 'On Periodicity and Radioactivity' (in Russian), Atomizdat, Moskva 1974.
[22] Wojtkowiak B., *Histoire de chimie*, Tech. et Doc.–Lavoisier, Paris 1984.

Particular References

[23] Aristotle, 'Works', from: 'Great Books of Western World', Vol. 8, Encyclopedia Britanica, Chicago 1952.
[24] Lucretius Carus T., *De Rerum Natura*, from: Munro H. A. J. Ed., Deighton Bell and Co., Cambridge 1866, Vol. II, p. 136 (Book II, 294–296).
[25] Bacon R., in: Ramsay W., *Nature*, 31 August, 282 (1911), from [5].
[26] Eliade M., 'The Forge and the Crucible', Harper and Row, New York 1971.
[27] Ploss E. I., Ed., *Alchimia*, Heinz Moss Verlag, München 1970.
[28] Valentinus Basilius, *Vier Traktätlein*, in: *Dyas Chimica Tripartita*, Luca Jennis, Franckfurt am Mayn 1625, p. 34.
[29] Maier M., *De Circulo Quadrato Physico Hoc est Auro*, Luca Jennis, Oppenheim 1616, p. 14.
[30] Sędziwój (Sendigovius) M., *Operatie Elexiris Philosophici*, 1586, from Polish transl. by R. Bugaj, PWN, Warszawa 1974.
[31] Libavius A., *Alchemia*, Francfourt 1597, from German transl., *Alchemia*, Chemie Verlag, Weinstein 1974.
[32] Boyle R., 'The Sceptical Chymist', 1661, from ed. by Dent J., London (without publication year).
[33] Galileo Galilei, *Sermones de motu gravium*, 1590, from *Le opere di Galileo Galilei*, Firenze 1854, Vol. 11, p. 29.
[34] Rogaliński J., 'Experience of Effects Perceived by the Senses' (in Polish), Book, I, Drukarnia J.K.M. Societatis Jesu, Poznań 1765.
[35] Lomonosov M. V., 'Complete Works' (in Russian), Izd. AN SSSR, Moskva 1950 (Vol. I), 1951 (Vols. II and III).
[36] Lavoisier A., *Traité élémentaire de chimie*, Paris 1801, Deterville Librairie, from [19].
[37] Kuznecova O. V., 'The Atomic Conception of the Structure of Substances in the 19th Century' (in Russian), Nauka, Moskva 1983.
[38] Ditfurth H. von, *Kinder des Weltalls*, Hoffmann und Campe Verlag, Hamburg 1970, *Am Anfang war der Wasserstoff*, Hoffmann und Campe

Verlag, Hamburg 1973, *Der Geist fiel nicht vom **Himmel***, Hoffmann und Campe Verlag, Hamburg 1976.
[39] Fleck L., *Entstehung und Entwicklung einer wissenschaftlichen Tatsache*, Benno Schwabe und Co., Basel 1935.
[40] Fuliński A., 'Science in the Context of Culture, in: Science, Religion, History' (in Polish). *Proceedings from a Seminar in Castel Gandolfo, August 1980*, Rome 1981.
[41] Solla Price D. J. de, 'Science since Babilon', Yale University Press, New Haven 1961.
[42] *Ein wahrhaftige Lehr der Philosophie von der Begehrung der Metallen und ihrem rechten Beginnen* (author unknown), in: *Dyas Chimica Tripartita*, Luca Jennis, Franckfurt am Mayn 1625.
[43] Paracelsus (Theophrastus Bombast von Hohenheim), from [19], p. 18.
[44] Paracelsus, *Von dem Saltz und von dem was Saltz enthält*, in: *Paracelsus sämtliche Werke in neuzeitliches Deutsch*, Gustaw Fischer, Jena 1930, Vol. III, p. 620.
[45] Tachenius O., *Hippocrates Chimicus*, Brunsvig 1668, p. 11.
[46] *Hermetico-Spagyrisches Lustgärtlein*, in: *Dyas Chimica Tripartita*, Luca Jennis, Franckfurt am Mayn 1625.
[47] Stahl G. E., *Zymotechnia Fundamentalis seu Fermentationis Teoria Generalis*, Halle 1697.
[48] Rey J., *Sur la recherche de la cause pour laquelle l'estain et le plomb augmentent de poids quand on les calcine*, 1630, from Alembic Club, Edinburgh 1893.
[49] Helmont J. van, *Ortus Medicinae*, L. Elsevivius, Amsterodami 1648, p. 109.
[50] Jungius J., *Dioscopiae physicae minores*, 1662, in: Kargo H., *Joachim Jungius Experimente zur Begründung der Chemie als Wissenschaft*, Franz Steiner Verlag, Wiesbaden 1968.
[51] Śniadecki J., 'The Principles of Chemistry' (in Polish), 3rd ed., ed. by J. Zawadzki, Wilno 1816 Vol. I, 1817 Vol. II.
[52] Becher J. J., *Actorum Laboratorii Chymici Monacensis, seu Physica Subterranea*, 1669; 3rd ed. by J. D. Zunner, Francfurt 1681.
[53] Lemery N., *Cours de chymie*, 1675, from: 9th ed., Estienne Michellet, Paris 1697 and new edition, Laurent-Charles D'Houry, Paris 1754.
[54] Nollet J. A. *Leçon de physique experimentale*, 1745, from: 3rd ed. by Frères Guerin, Paris 1754, Vol. IV, p. 163.
[55] Becher J. J., *Physica Subterranea*, 1667 from new edition, Lipsiae 1738.
[56] Stahl G. E., *Zufällige Gedanken und nützliche Bedenken über den Streit von sogenannten Sulphure, und zwar sowohl von einem gemeinen verbrennlichen oder flüchtigen als unverbrennlichen oder fixen*, Halle 1716, from [19], p. 60.
[57] Stahl G. E., *Fundamenta Chymiae Dogmaticae et Experimentalis*, 2nd ed., Johannes Isaacus gen. Hollandus, Norimbergae 1747, Part III, p. 391.

[58] Stahl G. E., *Spacimen Becherianum*, 1738, from [55], p. 148.
[59] Stahl G. E., *Billig Bedenken, Erinnerungen und Erläuterung über D. Bechers Naturkundigung der Metalle*, Christoph Multz, Franckfuhrt und Leipzig 1723, p. 70.
[60] Charderon P., *Digr. Acad. Dijon* (1769), from [16], Vol. III, p. 610.
[61] Guyton de Morveau L. B., *Digr. Acad. Dijon* (1772), from [16], Vol. III, p. 611.
[62] Hermbstädt S. F., *Versuche und Beobachtungen*, Berlin 1789, Vol. II, p. 177.
[63] Bugaj R., 'Sendigovius (1566–1636). Life and Work' (in Polish), Ossolineum, Wrocław 1968.
[64] Priestley J., 'Experiments and Observations on Different Kinds of Air', 1775, from Alembic Club, No 7, Edinburgh 1894.
[65] Scheele C. W., *Chemische Abhandlungen von der Luft und Feuer*, Uppsala–Leipzig 1777, p. 13.
[66] Lavoisier A. L., *Memoire sur la calcination d'etain dans les vaissaux fermes et sur la cause de l'augmentation du poids*, 1774, from: Lavoisier A. L., *Oeuvres*, Paris 1864, Vol. II, p. 105.
[67] Lavoisier A. L., *Memoire sur la combustion en general*, 1777, from: Lavoisier A. L., *Oeuvres*, Paris 1864, Vol. II, p. 225.
[68] Lavoisier A. L., *Considerations generales sur la nature des acides*, 1778, from: Lavoisier A. L., *Oeuvres*, Paris 1864, Vol. II, p. 248.
[69] Lavoisier A. L., *Reflections sur phlogiston*, 1783, from: Lavoisier A. L., *Oeuvres*, Paris 1864, Vol. II, p. 623.
[70] Cavendish H., 'Experiments on Air', *Phil. Trans.*, **74**, 119 (1784) from: Alembic Club No 3, Edinburgh 1893.
[71] Rumford B., 'An Inquiry Concerning the Source of the Heat which is Excited by Friction, read before Royal Society, January 25, 1798', from: 'The Completed Works of Count Rumford', G. E. Ellis, Boston 1870, Vol. I, p. 490.
[72] Herapath J., *Ann. Phil.*, 17, 274 (1821), from [4].
[73] Śniadecki J., 'The Principles of Chemistry' (in Polish), 1st ed., Wilno 1800, p. 45.
[74] Śniadecki J., 'An Explanation of Some Points Concerning the Science of Caloric', 1815, in: 'The Works of Jędrzej Śniadecki' (in Polish), ed. by M. Baliński, Warszawa 1840, Vol. III, p. 165.
[75] Śniadecki J., 'On Dissolution. A Treatise Sent to the Royal Society of the Friends of Science in Warsaw in May 1805' (in Polish), Wilno 1806, p. 45.
[76] Berzelius J. J., *Lehrbuch der Chemie*, Arnoldsche Buchhandlung, Dresden 1825.
[77] Epicurus, 'A Speech to Herodotus', from [14] p. 18.
[78] Simplicius of Cilice, *Physica*, from [14] p. 59.

REFERENCES 265

[79] Newton I., 'Opticks', 2nd ed., William and John Innys, London 1717, from edition: Dover, New York 1952.
[80] Roscoe H. E., Harden A., 'A New View on Dalton's Atomic Theory', MacMillan and Co., London 1896.
[81] Dalton J., *J. Natural Phil.*, **28**, 81 (1811), from [4].
[82] Richter J. B., *Anfangsgründe der Stöchiometrie oder Messkunde chemischen Elementen*, Hirschberg, Breslau, 1792, Vol. I, p. XXIX and 206 and from [8] p. 183.
[83] Proust L. J., *Ann. chim.*, **32**, 30 (1799), from [19].
[84] Petruska-Madej E., *Kwart. Hist. Nauki i Techniki*, **29**, 377 (1987).
[85] Proust L. J., *Journ. de phys.*, **63**, 668 (1806).
[86] Berthollet C. L., *Essai de statique chimique*, Vols. I, II, Firmin Didot, Paris 1803.
[87] Dalton J., 'A New System of Chemical Philosophy', Manchester 1808, from [80].
[88] Dalton J., *Mem. Lit. Phil. Soc. Manchester*, **1**, 271 (1803), in: Millington J. P., 'John Dalton', J. M. Dent and Co., London 1906, p. 81.
[89] Dalton J., *Ein nueus System des chemischen Theiles der Naturwissenschaft*, Dümmler, Berlin 1812, Vol. I, p. 247.
[90] Laurent A., *Methode de chimie*, Mallet-Bachelier, Paris 1854.
[91] Berzelius J. J., *Ann. Phil.*, **2**, 443 (1813), from [4].
[92] Maxwell J., *Nature*, 432 (1873), from [4].
[93] Whewell W., *Philosophy of Inductive Science*, **1**, 405 (1840), from [4] p. 121.
[94] Berzelius J. J., *Ann. Phil.*, **3**, 443 (1814) from [4].
[95] Gay-Lussac J., *Mem. Soc. d'Arceuil*, **2**, 207 (1809), from Alembic Club N° 4, Edinburgh 1893, pp. 8, 10.
[96] Wollaston W. H., *Phil. Trans.*, **1**, 2 (1814).
[97] Zdzitowiecki J. S., 'An Introductory Lecture on Chemistry' (in Polish), Warszawa 1850 Vol. I, 1851 Vol. II, printed by Drukarnia Józefa Ungera.
[98] Prout W., *Ann. Phil.*, **6**, 321 (1815), from [5].
[99] Prout W., *Ann. Phil.*, **7**, 111 (1816), from [5].
[100] Prout W., in: Dauberny C., 'On the Atomic Theory', Oxford 1831, pp. 129–133, from [5].
[101] Turner E., *Phil. Mag.*, **III 1**, 109 (1832), from [4].
[102] Berzelius J. J., *Am. J. Sci.*, **48**, 369 (1845), from [5].
[103] Berthelot M., *L'unité de la matière—Les multiples de l'hydrogene et les elements polymers*, Paris 1885, from [5].
[104] Liebig J., 'Familiar Letters on Chemistry', Taylor, Walton and Maberly, London 1844.
[105] Kekulé A., *Laboratory*, **1**, 303 (1867), from [4].
[106] Mendelejeff (Mendeleev) D., *Ann. Chem. Pharm.*, **VIII Suppl.** (1871), from Ostwalds Klassiker N° 68, Leipzig 1895, p. 99,

[107] Avogadro A., *J. Phys.*, **73**, 58 (1811), from Alembic Club N° 4, Edinburgh 1893, p. 31.
[108] Cannizzaro S., *Nuovo Cimento*, **7** (1858), from Ostwalds Klassikier N° 30, Leipzig 1913, pp. 3–43.
[109] Young T., *Phil. Trans.*, **99**, 159 (1809), from [5].
[110] Ampère A., *Ann. Phys. Chim.*, **1**, 295 (1816), from [5].
[111] Dumas J. B., *Ann. Chim.*, (1851), from [16], Vol. IV, p. 884.
[112] Cooke J. P., *Am. J. Sci.*, **17**, 387 (1851), from [5].
[113] Chancourtois A. A. B. de, *Vis tellurique*, Paris 1863, from [5].
[114] Newlands J. A. R., *Chem. News*, **9**, 59 (1864), from [5].
[115] Odling W., *Quart. J. Sci.*, **1**, 642 (1864), from [5].
[116] Mendelejeff (Mendeleev) D., *Z. Chem.*, **12**, 405 (1869), from [5].
[117] Seubert K., in: Ostwalds Klassikier N° 68, Leipzig 1895, p. 134.
[118] Mendelejeff (Mendeleev) D., *J. Chem. Soc.*, **55**, 634 (1899), from Dąbkowska M., *Wiad. Chem.*, **38**, 779 (1983).
[119] Ostwald W., 'Faraday Lecture', *J. Chem. Soc.*, **85**, 520 (1904).
[120] Teske A., 'Marian Smoluchowski' (in Polish), PWN, Warszawa 1955, p. 169.
[121] Perrin J., *Comp. Rendus (Paris)*, **146**, 967 (1908).
[122] Perrin J., *Les atomes*, Alcan, Paris 1913, p. 65.
[123] Kurnatov N. S., from: Anosov W. J., Pogodin S. A., 'An Introduction to Chemical Analysis' (in Russian), Izd. AN SSSR, Moskva 1947.
[124] Zintl E., Goubeau, Bullenkopf, *Z. phys. Chem.*, A **154**, 1 (1931); Zintl E., Kaiser H., *Z. anorg. Chem.*, **211**, 113 (1933).
[125] Rutherford E., *Phil. Mag.*, **21**, 669 (1911).
[126] Rutherford E., *Proc. Roy. Soc. (London)*, A **97**, 374 (1920).
[127] Guyton de Morveau L. B., *Observation sur la physique*, **19**, 370 (1782).
[128] Guyton de Morveau L. B., Lavoisier A. L., Berthollet C. L., Fourcroy A. de, *Methode de nomenclature chimique*, Cuchet, Paris 1787.
[129] Czyrniański E., 'Polish Chemical Terminology' (in Polish), printed by Drukarnia Czasu, Kraków 1853.
[130] Snelders H. A. M., Spronsen J. W. van, *Scheets der Leere van Lavoisier door Martinus van Marum*, Koninklijke Nederlandse Chemische Vereniging, Den Haag 1987.
[131] Berzelius J. J., *Essai sur la theorie des proportions chimiques et sur l'influence chimique de l'electricite*, Mequignon–Marvis, Paris 1819, p. 73.
[132] Berzelius J. J., *Lehrbuch der Chemie*, Arnoldische Buchhandlung, Dresden 1827, Vol. III, p. 79.
[133] Helmholtz H., *J. Chem. Soc.*, **222**, 277 (1881).
[134] Berzelius J. J., *Jahresbericht*, **11**, 44 (1832), from [19].
[135] Mitscherlich E., *Ann. Pharm.*, **12**, 306 (1834).
[136] Dumas J. B., Liebig J., *Comp. Rendus (Paris)*, **5**, 567 (1837), from [19].

REFERENCES

[137] Liebig J., *Ann. Chem.*, **26**, 113 (1838).
[138] Dumas J. B., *Comp. Rendus (Paris)*, **9**, 813 (1839), from [19].
[139] Dumas J. B., *Ann. chim. phys.*, **73**, 73 (1840).
[140] Gerhardt Ch., *Traité de chimie organique*, F. Didot, Paris 1853–1856.
[141] Williamson A., *Ann. Chem.*, **77**, 87 (1851).
[142] Frankland E., *Phil. Trans.*, 417 (1852).
[143] Kekulé A., *Ann. Chem.*, **104**, 133 (1857).
[144] Kekulé A., *Ann. Chem.*, **106**, 155 (1858).
[145] Kekulé A., *Lehrbuch der organischen Chemie*, 2nd ed., Vol. I, F. Enke, Erlangen 1867, p. 92.
[146] Kolbe H. W., *Ausführliches Lehrbuch der organischen Chemie*, Vieweg, Braunschweig 1854–1859.
[147] Butlerov A., *Z. Chem. Pharm.*, 594 (1861).
[148] Butlerov A., *Zh. Rus. Khim. Obshch.*, **11**, 289 (1879).
[149] Butlerov A., *Z. Chem. Pharm.*, 500 (1863).
[150] Mierzecki R., Kokowska J., *Wiad. Chem.*, **37**, 263 (1983).
[151] Kekulé A., *Sur la constitution des substances aromatique*, Bull. Soc. Chim. Fr. N.S., **3**, 98–100 (1865); also in: [19].
[152] Kekulé A., *Chemie der Benolderivate*, F. Enke, Erlangen 1867.
[153] Pasteur L., *Comp. Rendus (Paris)*, **26**, 535 (1848); *Ann. chim. phys.*, **24**, 442 (1848), from *Oeuvres de Pasteur*, Masson, Paris, 1922, Vol. I, pp. 61–80.
[154] Pasteur L., *Recherche sur la dissimetrie moleculaire des produits organiques naturelles*, Paris 1860, from *Oeuvres de Pasteur*, Masson, Paris 1922, Vol. I, p. 327.
[155] Hoff J. H. van't, *La chimie dans l'espace*, P. M. Bazendijk, Rotterdam 1875.
[156] Baeyer A., *Ber.*, **8**, 2278 (1885).
[157] Mierzecki R., *Wiad. Chem.*, **6**, 214 (1952).
[158] Clausius R., *Ann. Phys.*, **101**, 347 (1857).
[159] Ostwald W., *J. prakt. Chem.*, **30**, 93, 225 (1884); **31**, 433 (1885); **32**, 300 (1885); **33**, 352 (1886); *Z. phys. Chem.*, **1**, 74, 96 (1887).
[160] Arrhenius S., *Biehang der Stockholmer Akademie*, **8**, 13, 14 (1884).
[161] Hoff J. H. van't, *Etudes de dynamique chimique*, Federic Muller and Co, Amsterdam 1884.
[162] Hoff J. H. van't, *Z. phys. Chem.*, **1**, 481 (1887).
[163] Hoff J. H. van't, 'Nobel Lecture' (1901), from 'Nobel Lectures Chemistry', Elsevier, Amsterdam 1966.
[164] Nernst W., *Z. phys. Chem.*, **4**, 129, (1889).
[165] Arrhenius S., *Z. phys. Chem.*, **1**, 631 (1887).
[166] Ostwald W., *J. prakt. Chem.*, **32**, 300 (1885).
[167] Ostwald W., Nernst W., *Z. phys. Chem.*, **3**, 120 (1889).
[168] Czyrniański E., 'Theory of Chemical Compounds' Formation on the

Basis of the Rotational Motion of the Atoms' (in Polish), Uniwersytet Jagielloński, Kraków 1862.
[169] Johnson O. C., *Chem. News*, **42**, 51 (1880).
[170] Werner A., *Vierteljahrschrift Naturforsch. Ges. (Zürich)*, **36**, 129 (1891), from [19].
[171] Werner A., *Z. anorg. Chem.*, **3**, 267 (1893); *Ann. Chem.*, **322**, 261 (1902).
[172] Abegg R., Bodländer G., *Z. anorg. Chem.*, **20**, 453 (1899).
[173] Abegg R., *Z. anorg. Chem.*, **39**, 330 (1904).
[174] Falk K. G., Nelson J. M., *J. Am. Chem. Soc.*, **32**, 1637 (1910).
[175] Dajn G., *Zh. Rus. Khim. Obshch.*, **46**, 815 (1914).
[176] Thomson J. J., *Phil. Mag.*, **27**, 757 (1914).
[177] Mierzecki R., *Człowiek i Światopogląd*, **10**, 61 (1978).
[178] Lewis G. N., *J. Am. Chem. Soc.*, **38**, 762 (1916).
[179] Langmuir I., *J. Am. Chem. Soc.*, **41**, 868 (1919).
[180] Lewis G. N., 'Valence and Structure of Atoms and Molecules', New York 1923, from [4].
[181] Lewis G. N., *Trans. Farad. Soc.*, **19**, 453 (1923).
[182] Uhlenbeck G., Goudsmit S., *Naturwiss.*, **13**, 953 (1925).
[183] Pauli W., *Z. Phys.*, **31**, 765 (1925).
[184] Boerhaave H., 'A textbook written furtively from student's note-book'. 1724, from [19], p. 64.
[185] Albertus Magnus, from [8], p. 96.
[186] Geoffroy E. F., *Mem. Acad. Sci. (Paris)*, 212 (1718), from [19], p. 68.
[187] Bergman T., *De Attractionis Electivis*, 1775, in: Bergman T., *Opuscula Physica et Chemica*, Lipsiae 1780, Vol. III, from [19].
[188] Lavoisier A. L., Laplace P., *Memoire sur la chaleur*, Paris 1783.
[189] Berthollet C. L. *Recherches sur les lois d'affinite*, from [19], p. 193.
[190] Śniadecki J., "A Theory of Organic Entities' (in Polish), printed by Drukarnia No 646, Warszawa 1804.
[191] Berthelot M., *Chimie organique fondée sur la synthese*, Mallet–Bachelter, Paris 1860, Vol. II, p. 805.
[192] Berzelius J. J., *Ann. chim.*, **61**, 146 (1836), from [19].
[193] Williamson A. W., *Phil. Mag.*, **37**, 550 (1851).
[194] Wilhelmy L., *Ann. Phys. Chem.*, **21**, 413 (1850).
[195] Williamson A. W., *Ann. Chem.*, **81**, 73 (1852); **91**, 236 (1854).
[196] Berthelot M., Saint-Giles L. P. de, *Ann. chim. phys.*, **66**, 5 (1863).
[197] Berthelot M., *Ann. chim. phys.*, **66**, 110 (1863).
[198] Boguski J. J., *Ber.*, **9**, 6 (1876); *Kosmos*, **1**, 258 (1876).
[199] Berthelot M., Saint-Giles L. P. de, *Ann. chim. phys.*, **68**, 225 (1864).
[200] Guldberg C. M., Waage P., *Etudes sur les affinites chimiques*, 1867, from Russian translation, ed. by Adolf Darre, Kharkov, 1892.
[201] Guldberg C. M., Waage P., *J. prakt. Chem.*, **19**, 69 (1879).
[202] Le Châtelier H. L., *Comp. Rendus (Paris)*, **93**, 786 (1884).

[203] Hess G., *Ann. chim. phys.*, **75**, 80, 325 (1840).
[204] Thomsen J., *Grundzüge eines thermochemischen System*, in: *Pogg Ann.*, **92**, 35 (1854).
[205] Berthelot M., *Essai de mechanique chimique fondée sur la thermochimie*, Dunod, Paris 1879, p. XXVIII.
[206] Mierzecki R., *Wiad. Chem.*, **36**, 305 (1982).
[207] Horstmann A., *Ber.*, **2**, 137 (1869); **4**, 635 (1871).
[208] Liebig J., *Ann. Chem.*, **30**, 254 (1839).
[209] Ostwald W., Kataliz, Moscow 1903, p. 19; in: [15], p. 267.
[210] Zawidzki J., *Bull. Acad. Cracovie*, **A**, 275, 379 (1916); 'Chemical Kinetics' (in Polish), Komisja Uczczenia Pamięci Prof. Jana Zawidzkiego, Warszawa 1931, p. 103.

Index of Names

Abu Ali al Hasayn ibn Abd Allah ibn Sina (980–1037), 7, 249
Abegg Richard (1869–1919), 199, 200, 218
Adelhard of Bath (12th c.), 8, 97
Adet Auguste Pierre (1763–1832), 122
Agricola, see Bauer Georg
Albertus Magnus of Bollstädt (1193–1280), 8, 38, 212, 250, 268
Alexander of Aphrodisias (2nd c.), 96
Alembert Jean le Rond d' (1717–1783), 220
Ampère André Marie (1775–1836), 131, 132, 136, 137, 266
Anaxagoras of Clathomen c. 500–c. 428 BC, 28
Anaximenes of Miletus (c. 585–c. 525 BC), 28, 248
Anderson Carl David (1905–1974), 157
Arnold de Villanova (c. 1235–c. 1311), 12, 250
Archimedes (287–212 BC), 248
Aristotle (384–322 BC), 3, 4, 5, 8, 28, 36, 37, 54, 64, 67, 90–97, 159, 207–210, 224, 246, 248, 262
Arrhenius Svante August (1859–1927), 194, 195, 242, 245, 256, 267
Avicenna, see Abu Ali
Avogadro de Quarenga Amadeo (1776–1856), 106, 131, 132, 266

Bacon Francis (1561–1626), 17
Bacon Roger (c. 1214–c. 1294), 8, 18, 247, 262
Baeyer Adolf (1835–1917), 190, 191, 267
Barton Derek Harold (1918–), 193
Basilius Valentinus? (1394–), 262
Bauer Georg (1434–1541), 250
Baumé Antoine (1728–1804), 161

Beaumont Elie (1798–1879), 139
Becher Johann Joachim (1635–1682), 59, 63, 207, 251, 263
Becquerel Antoine Henri (1852–1908), 152
Becker J. (20th c.), 156
Bergman Thorbjörn Olaf (1735–1782), 83, 219, 220, 223, 268
Bernal John Desmont (1901–), 66, 247, 261
Berthelot Marcelin Pierre Eugène (1827–1907), 41, 128, 177, 228, 231, 236, 237, 239, 255, 265, 268, 269
Berthollet Claude Luis (1748–1822), 108, 109, 149, 150, 161, 168, 221–224, 231, 252, 265, 268
Berzelius Jöns Jacob (1770–1848), 90, 92, 117, 120–122, 124, 125, 127, 132, 163–171, 177, 194, 223, 229, 240, 250, 253, 264–266, 268
Biot Jean Baptiste (1774–1862), 187
Bischoff Carl (1855–1908), 191, 193
Black Joseph (1757–1827), 63, 68, 82
Blackett Patrick Maynard (1897–1979), 156
Bodländer Guido (1855–1904), 199, 268
Boerhaave Herman (1668–1738), 50, 62, 81, 211, 224, 252, 268
Boguski József Jerzy (1853–1933), 231, 268
Bohr Niels (1885–1962), 201, 203, 259
Boisboudran Lecoq de (1838–1912), 144
Bolos of Mendes (2nd c. BC), 10, 248
Boltzmann Ludwig (1844–1906), 147, 148, 238
Bombast von Hohenheim Theophrastus (Paracelsus) (1493–1541), 12, 13, 38, 43–46, 54, 57, 263
Borelli Giovani Alphonso (1608–1679), 98
Boscovic Rudjer Josip (1711–1787), 100

INDEX OF NAMES

Bothe Walter (1891–1957), 156
Boyle Robert (1627–1691), 3, 16, 18, 19, 50–58, 61, 62, 64, 66, 76, 78, 98, 99, 136, 160, 213–215, 251, 262
Bridgman Percy William (1882–1961), 24
Broek Jan Atram van den (20th c.), 154
Bruno Giordano (1548–1600), 97
Buff Heinrich (1828–1872), 177
Buffon Georges Leclerc (1707–1788), 63
Bugaj Roman (20th c.) 69, 264
Butlerov Aleksandr (1828–1886), 175, 180, 181, 186, 205, 236, 255, 267

Cannizzaro Stanislao (1826–1910), 133, 134, 255, 266
Cardano Gerolamo (1501–1576), 49
Carlisle Antony (1768–1840), 163
Carnot Nicolas Leonard Sadi (1796–1832), 238
Cavendish Henry (1731–1810), 70, 73, 77, 106, 264
Celsus Aulus Cirnelius (53 BC–7), 248
Centnerszwer Mieczysław (1874–1944), 229
Chadwick James (1891–1974), 156
Chancourtois Alexandre Emile Beguyer de (1819–1886), 139, 140, 266
Charderon Jean Pierre (1714–1769), 67, 264
Clausius Rudolf Emanuel (1822–1888), 134, 194, 238, 243, 255, 267
Cooke Josiah Parson (1827–1894), 138, 266
Copernicus, see Kopernik
Couper Archibald Scott (1831–1892), 179
Crookes William (1832–1919), 138, 266
Curie Maria b. Skłodowska (1867–1934), 145, 153, 255, 257
Curie Pierre (1859–1906), 153, 257
Czyrniański Emilian (1824–1888), 162, 182, 183, 196, 266, 267

Dajn S. W. (20th c.), 200, 202, 268
Dalton John (1766–1844), 83, 100, 106, 109–122, 124, 125, 127, 166, 206, 223, 253, 265

Davy Humphry Bartholomew (1778–1829), 163, 164, 166, 253
Davy John (1790–1868), 169
Dąbrowski Stefan (1877–1947), 149
Debye Peter (1884–1966), 200, 258
Deiman Jan Rudolf (1743–1808), 163, 169
Democritus (c. 460–c. 370 BC), 5, 8, 29, 61, 93, 94, 95, 97, 101, 159, 207, 248
Dirac Paul Adrien (1902–1984), 204
Ditfurth Hoimar (1921–), 25, 262
Döbereiner Johann Wolfgang (1780–1849), 137, 138, 141
Duhem Pierre (1861–1916), 238
Dulong Pierre Louis (1785–1838), 127
Dumas Jean Baptiste André (1800–1884), 128, 133, 137, 169–171, 173, 253, 266, 267

Einstein Albert (1879–1955), 23, 147, 148, 258
Eliade Mircea (1907–1986), 10, 262
Empedocles of Akragas (c. 480–c. 430 BC), 29, 37, 248
Epicurus (341–270 BC), 96, 98, 101, 264
Ercker Lasarus von Schreckenfels (c. 1530–1594), 250
Erlenmeyer Emil (1825–1909), 183
Erxleben Johann Christian Policarp (1744–1777), 63
Euclid (365–300 BC), 5
Euler Leonard (1707–1783), 20

Fajans Kazimierz (1887–1975), 154, 259
Faraday Michael (1791–1867), 166, 167, 169, 194, 253
Fleck Ludwik (1896–1961), 25, 263
Fourcroy Antoine François (1755–1809), 161
Fuliński Andrzej (20th c.), 25, 263

Galen Claudius (130–200), 6, 249
Galileo Galilei (1564–1642), 17, 42, 49, 220, 251, 262
Galvani Luigi (1737–1798), 163
Gassendi (Gassend) Pierre (1592–1655), 97, 102, 106, 221

INDEX OF NAMES

Gay-Lussac Louis Joseph (1778–1850), 114, 124, 126, 127, 131, 149, 167, 265
Geber, see Jabir ibn Hajjan
Geiger Hans (1882–1945), 153
Gell-Mann Murray (1929–), 158
Geoffroy François Etienne (1672–1731), 136, 217, 218, 252, 268
Gerhardt Charles (1816–1856), 133, 172–175, 179, 267
Gibbs Josiah Willard (1839–1903), 238, 240
Gilbert William (1540–1611), 250
Giua Michel (1889–1966), 198, 261
Gladstone John (1827–1902), 128, 138
Glauber Johann Rudolf (1604–1668), 169, 213, 251
Gmelin Leopold (1788–1853), 138
Goethe Johann Wolfgang (1749–1832), 10
Goldstein Eugen (1850–1930), 151
Goudsmit Samuel Abraham (1902–1978), 155, 203, 268
Grossereste Robert (1175–1253), 8, 9, 18
Grotthus Theodor Christian (1785–1822), 66, 166
Guldberg Otto Maximilian (1836–1902), 232, 234, 240, 256, 268
Guyton de Morveau Louis Bernard (1737–1816), 67, 161, 262, 266

Hantsch Artur Rudolf (1857–1935), 187
Hapelius Nicolaus (1559–1622), 64
Harkins William Draper (1873–1951), 154
Hassel Odd (1897–1981), 193
Hassenfrantz Jean André (1755–1827), 122
Heisenberg Werner Carl (1901–1976), 24, 91, 156, 204, 259
Heitler Walter Heinrich (1904–1981), 203, 204
Helmholtz Hermann Ludwig (1821–1894), 89, 166, 194, 202, 253, 256, 266
Helmont Jan Baptist van (1577–1644), 50, 51, 69, 70, 225, 250, 263
Heraclitus of Ephesus (c. 540–c. 480 BC), 2, 29, 47, 248

Herapath J. (1790–1868), 82, 264
Hermbstädt Sigismundus Friedrich (1760–1833), 68, 215, 264
Herschel John Frederic William (1792–1871), 22, 82, 187
Hess Germain Henri (1802–1850), 235, 269
Hevesy György (George) de (1885–1966), 154
Hinshelwood Cyril Norman (1897–1967), 238
Hippocrates (460–377 BC), 248
Hoff Jacobus Hendricus van't (1852–1911), 175, 188–191, 195, 196, 202, 234, 235, 239, 256, 267
Hofmann August Wilhelm (1818–1892), 173
Hooke Robert (1635–1703), 71
Horstmann August Friedrich (1842–1929), 240, 269
Huygens Christian (1629–1695), 250
Humboldt Friedrich Heinrich Alexander Freiherr von (1769–1859), 114
Hund Friedrich (1896–), 204
Hunt T. Sterry (1826–1852), 128

Ivanenko Dimitr (1904–), 156

Jabir ibn Hajjan (Geber) (c. 720–c. 813), 7, 37, 47, 249
Jaśkiewicz Jan Dominik (1749–1809), 84
Johnson Otis Coe (2nd half of 19th c.), 197, 268
Joliot-Curie Frédéric (1900–1958), 96, 259
Joliot-Curie Irène (1893–1956), 96, 259
Joule James Prescott (1818–1889), 89
Jungius Jung Joachim (1587–1657), 51, 98, 263

Keesom Willem Hendrik (1876–1956), 200
Kekulé August (1829–1896), 128, 175, 177, 178, 181–186, 196, 205, 255, 256, 262, 267
Kipping Frederic Stanley (1863–1949), 191
Kirwan Richard (1735–1812), 77
Kohlrausch Friedrich Wilhelm Georg (1840–1910), 194
Kohlrausch K. W. F. (20th c.), 192

Kolbe Adolf Wilhelm Hermann (1818–1884), 175, 179, 187, 227, 267
Kopernik (Copernicus) Mikołaj (1473–1543), 16, 250
Kröner Wilhelm (1839–1925), 185
Krönig August Karl (1822–1879), 134
Kurnakov Nicolai S. (1860–1941), 150, 266
Kuznetsova Olga V. (20th c.), 23, 262

Landenburg Albert (1842–1911), 190
Langmuir Irwing (1881–1957), 202, 268
Laplace (La Place) Pierre Simon de (1749–1827), 220, 221, 223, 235–237, 239, 266
Laurent Auguste (1807–1853), 117, 133, 169, 171–173, 185, 265
Lavoisier Antoine Laurent (1743–1794), 18, 20, 56, 65, 69, 72–81, 83, 84, 90, 105, 106, 110, 136, 160, 162, 171, 206, 220, 221, 223, 235–237, 239, 252, 262, 264, 268
Le Bel Achile Jacques (1847–1930), 188, 156
Leblanc Nicolas (1742–1806), 66, 252
Le Chatelier Henri (1850–1930), 234, 268
Lemery (L'Emery, Le Mery) Nicolas (1645–1715), 59–62, 68, 215, 250, 263
Leucippus of Miletus (c. 490–420 BC), 61, 93
Lewis Gilbert Newton (1875–1946), 201–203, 258, 268
Libau (Libavius) Andreas (1540–1615), 13, 46, 47, 250, 262
Liebig Justus (1803–1873), 129, 132, 167, 169, 170, 227, 228, 240, 242, 253, 265–267, 269
Lockeyer Joseph Norman (1836–1920), 128
Lomonosov Mikhail (1711–1765), 19, 20, 22, 50, 62, 81, 100, 101, 252, 262
London Frantz (1900–1954), 203, 204
Lorentz Hendrik Anton (1853–1920), 152
Loschmidt Joseph (1821–1895), 145, 256
Lowitz Johann Tobias (1757–1804), 168
Lucretius Titus Carus (c. 95–c. 55 BC), 3, 19, 93, 248, 262
Lullus Raimundus (1235–1315), 250

Mach Ernst (1838–1916), 146

Macquer Pierre Joseph (1718–1784), 73, 161
Maier Michael (1568–1662), 11, 214, 262
Malus Etienne Louis (1775–1812), 186
Maxwell James Clerk (1831–1879), 89, 121, 147, 265
Mayow John (1645–1679), 71
Meitner Lise (1878–1968), 156
Mendeleev Dimitr Ivanovich (1834–1907), 130, 131, 141–144, 154, 199, 255, 265, 266
Meyer Edgar (20th c.), 153
Meyer Julius Lothar (1830–1895), 141
Mierzecka Anna (1919–1970), 247
Mierzecki Roman (1921–), 267, 268, 269
Mitscherlich Eilhardt (1794–1863), 127, 168, 266
Mohr E. (20th c.), 193
Moses (13th c. BC), 98
Mylius Daniel Johann (17th c.), 47

Natanson Władysław (1864–1937), 243
Neddermeyer Seth (1907–), 157
Nernst Walter Hermann (1864–1941), 195, 196, 257, 258, 267
Newlands John Alexander (1838–1898), 140, 256, 266
Newton Isaac (1643–1727), 18, 19, 79, 99, 100, 105, 117, 118, 206, 216, 251, 265
Nicolas of Autrecourt (c. 1300–1350), 97
Nicolas of Cusa (1401–1464), 97
Nicolson William (1753–1815), 163
Nilson Lars (1840–1899), 144
Nollet Jean Antoine (1700–1770), 62, 82, 263

Occhialini Giuseppo Paulo Stanislao (1907–), 156
Ockham (Occam) William (1300–1349), 8, 250
Odling William (1829–1921), 140, 141, 175, 266
Olympiodoros (5th c.), 49
Ostwald Wilhelm (1853–1932), 146–148, 194, 196, 197, 240, 255, 256, 261, 266, 267, 269

INDEX OF NAMES

Palissy Bernard (1499–1589), 213
Paracelsus, see Bombast
Parmenides of Elea (c. 540–c. 470 BC), 3, 19
Partington James Riddick (1886–1965), 67, 261
Pascal Blaise (1623–1662), 17
Pasteur Louis (1822–1895), 187, 267
Paterno Emanuele (1847–1936), 188
Pauli Wolfgang (1900–1958), 203, 268
Pauling Linus Carl (1901–), 166, 204, 205, 259
Pean de Saint Giles Louis (1832–1863), 231, 268
Perrin Jean Baptiste (1870–1942), 149, 152, 266
Petit Alexis Thérèse (1791–1820), 127
Petruska-Madej Elżbieta (20th c.), 107, 265
Pettenkoffer Max Joseph (1818–1901), 138
Philipp K. (20th c.), 156
Pitzer Kennet Sandborn (1914–), 192
Planck Max Carl Ernst (1858–1947), 23, 258
Plato (427–347 BC), 4, 29, 95, 248
Plutarch (50–125), 28
Poincaré Henri (1854–1912), 22
Pope William Jackson (1870–1939), 191
Popper Karl Raimund (1902–), 24
Powell Cecil Frank (1903–1969), 157, 205
Priestley Joseph (1733–1804), 70–73, 77, 81, 252, 264
Prigogine Ilya (1917–), 244
Proust Louis Joseph (1754–1826), 107, 108, 125, 265
Prout William (1785–1850), 126–128, 265
Ptolemy (100–168), 5, 249
Pythagoras of Samos (c. 572–c. 497 BC), 4, 248

Ramsay William (1852–1916), 202
Rasseti F. (20th c.), 155
Regnault Henri Victor (1810–1878), 169
Remane Horst (20th c.), X, 262
Rey Jean (1583–1645), 18, 49, 50, 61, 62, 263

Richter Jeremias Beniamin (1762–1807), 106, 107, 125, 265
Rogaliński Józef (1728–1802), 19, 100–105, 262
Röntgen Wilhelm Conrad (1845–1923), 152
Rubinowicz Wojciech (Adalbert) (1889–1962), 155
Rumford, see Thomson Beniamin
Russel Alexander Smith (1888–1972), 154
Rutherford Daniel (1749–1819), 70
Rutherford Ernest (1871–1937), 153, 155, 156, 203, 258, 259, 266

Scheele Karl Wilhelm (1742–1786), 61, 66, 76, 81, 168, 187, 264
Scheidt Franciszek (1759–1807), 84
Schrödinger Erwin (1887–1961), 24, 204, 259
Semenov Nicolai (1896–), 242
Sendigovius (Sędziwój) Michał (1566–1636), 12, 39, 69, 71, 262
Sennert Daniel (1572–1637), 48, 57, 64, 251
Sidgwick Nevil Vincent (1873–1952), 205
Simplikios of Cilice (?–549), 96, 264
Slater John Clark (1900–1978), 204
Smoluchowski Marian (1872–1917), 147, 148, 258
Soddy Federic (1877–1956), 153, 154
Solla Price Derek de (?–1983), 26, 263
Sommerfeld Arnold (1868–1951), 203
Stahl Georg Ernst (1660–1734), 48, 63–65, 215, 252, 263, 264
Stas Jean Serge (1813–1891), 173
Stewart (20th c.), 154
Stolz Rüdiger (20th c.), X, 262
Stoney George Johnstone (1826–1911), 152, 257
Strube Irene (20th c.), X, 262
Strube W. (20th c.), 241
Suchten Alexander, see Zuchta
Svedberg Theodor (1884–1971), 148, 258
Sylvius de la Boe Francisc (1614–1672), 213
Szabadvary Ferenc (20th c.), 66

INDEX OF NAMES

Śniadecki Jędrzej (1768–1838), 56, 84, 88–90, 105, 106, 110, 217, 225, 226, 244, 245, 253, 263, 264, 268

Tachenius (Tache) Otto (1620–1690), 46, 135, 263
Tamm Igor (1895–1971), 156
Thales of Miletus (c. 620–c. 540 BC), 2, 28, 248
Thiele Johann (1865–1918), 186
Thomas Aquinas (1225–1274), 8, 38, 47
Thompson Beniamin Rumford (1753–1814), 81, 82, 87, 221, 252, 264
Thomsen Julius (1826–1909), 236, 237, 239, 269
Thomson Joseph John (1856–1940), 152, 153, 199, 200, 201, 268
Thomson Thomas (1773–1852), 125
Thomson William (Lord Kelvin) (1824–1907), 243
Torricelli Evangelista (1608–1647), 17
Troostvijk Adriaan Paets (1752–1836), 163
Turner Edward (1798–1837), 128, 265

Uhlenbeck George Eugene (1900–1977), 155, 203, 268

Volta Allessandro (1745–1827), 163, 253

Waage Peter (1833–1900), 232–234, 240, 256, 268

Waals Johannes Diderik van der (1837–1923), 255
Wallerius Jean Gottaschlak (1709–1785), 68
Werner Alfred (1866–1919), 150, 191, 197, 198, 205, 268
Whewell William (1794–1866), 22, 121, 260
Wien Wilhelm (1864–1928), 152
Wilhelmy Ludwig (1812–1864), 230, 268
Williamson Alexander Wilhelm (1824–1904), 173, 174, 230, 267
Winkler Klemens (1838–1904), 144
Wislicenus Johannes (1835–1902), 191
Wöhler Friedrich (1800–1882), 167, 169, 227, 253
Wojtkowiak Bruno (20th c.), X, 262
Wollaston William Hyde (1766–1828), 107, 124, 125, 265

Young Thomas (1773–1829), 89, 136, 220, 266
Yukawa Hideki (1907–1981), 157

Zahn K. (20th c.), 192
Zawidzki Jan Wiktor (1866–1928), 242, 269
Zdzitowiecki Józef Seweryn (1802–1879), 126, 265
Zintl E. (20th c.), 150, 266
Zosimos of Panopolis (c. 350–420), 6, 12, 249
Zuchta (von Suchten) Aleksander (1520–1590), 48

Index of Subjects

ability (capacity) to action, 34, 160, 206, 207–245
ab initio calculations, 26
academies and scientific societies, 21, 248, 249, 250
activation energy, 241, 242, 245
active properties, elements, 32, 33, 60, 210
activity coefficient, 233
affinity, 105, 114, 136, 171, 175, 178, 180, 189, 198, 200, 206, 209, 212, 213, 215, 217, 219–224, 226, 228–230, 233, 237, 245
   elective, 219, 222, 252
   law of, 109
   standard, 240
   table of, 217, 218, 252
   unit of (single bond), 177, 178, 180, 183–185, 236
air, as element, 28, 32–36, 47, 48, 69, 96, 209, 210, 248
   atmospheric, 72, 105
   dephlogisticated (purest, oxygen), 71, 73, 77
   fixed (carbon dioxide), 70
   inflamable (hydrogen), 66, 70, 78
   phlogisticated (nitrogen), 70, 77
alchemy, alchemists (spagirists), 7–16, 25, 36–47, 51, 52, 54, 79, 135, 211, 212, 246, 248, 249
Alexandria, Museum, Library, 6, 248
alkali, basic compounds, 10, 46, 135, 136, 144, 162, 165, 196, 258
alloys, 14, 105, 150, 214
anima, 15
anode, 166
antiparticle, 157, 158
aqua fortis, 48, 104

aqua regia, 213
aromatic compounds, 21, 184, 185, 206
asymetric carbon atom, 189, 190, 256
atom, as centre of vortical motion, 100
   as indivisible, 29, 93, 94, 96, 101, 117, 120, 122, 127
   as the smallest part, 19, 25, 30
   chemical, 129
   concept of, 91, 94, 106, 115–117, 119, 132, 133, 139, 145, 171
   definition of, 121
   model of, 153, 159, 202, 258
   parts of, 25, 102, 127–131
   physical, 129
   properties of, 101, 166, 191
   shape of, 95, 121, 159, 220
   symbols of, see **element**
atomic formulae, 115, 126, 180, 181
atomic heat, 127
atomic mass (weight), 91, 111–116, 124, 125, 126, 128, 133, 134, 136, 138–141, 147, 154, 176, 260
atomic theory, 96, 97, 116, 117, 121, 122, 124, 127, 132, 147, 248
atomic-molecular conception (hypothesis), 21, 29, 95, 98, 137
Avogadro constant, 149
Avogadro theory (hypothesis), 131–134, 253, 255
axiomatic reasoning, 5, 18, 24

basicity, see alkalis
benzene, discovery of, 169, 257
   structure of, 184, 190
bertholides, 149–150

INDEX OF SUBJECTS

bond, double, 174, 183, 186, 190, 201, 255
  coordinate, 205, 206
  covalent, 202
  energy of, 221
  ionic, 194, 197, 202
  mechanism of formation, 205
  triple, 190
Bronze Age, 1, 247
Brownian motions, 147, 148
burning, see combustion

calcination (roasting), 49, 61, 62, 65, 67, 73, 76
caloric (matter of heat), 68, 69, 83, 86–88, 105, 215, 216, 221, 252
calx, 50
canal rays, 152
caput mortuum (dead head), 60
causal determinant, 5, 208
catalysis, 230, 241–243, 253
cathode, 166
cathode rays, 151, 152
chain reaction, see reaction
chemical, force, see force
  formula, 113, 132
  groups, 138
  industry, 22, 67
  kinetics, 231, 232, 242, 257
  potential, 240
  resonance, 206, 259
  series, 172, 174
  symbols, see symbols
  types, see types
chemistry, inorganic (mineral), 21, 22, 26, 170, 228
  modern trends, 26
  organic, 22, 26, 133, 169, 170, 180, 188, 228
  orygin of the name, 6
  physical, 22, 26, 145
  quantum, 204, 205, 259
chlorination, 169, 170
chloroform, 169, 170

classification (systematization) of elements, see element of substances, 135, 136, 224
cold as principal quality, 32, 33, 36, 39, 209
combustibility, 38, 45, 59, 66, 67, 214, 215
combustion (burning), 48, 49, 54, 62–64, 71–73, 252
compound and mixture (solution), 96, 99, 108, 109, 162, 207, 208, 221
  complex, 196, 197
  non-polar, 200
  paired, 171
  polar, 200
conformational analysis, 191, 193
Congress International, 138, 159, 255
continuity of matter, 93, 96, 159
contraries, (contrarities), 11, 14, 30–32, 38, 94, 207–209, 211
coordinate bond, see bond
coordinate number, 198, 199
corpuscular theory (discontinuity of matter), 18, 21, 23, 81, 93, 97, 99, 100, 106, 109, 116, 117, 145, 148, 213, 215, 251

daltonides, 149–151
deduction, see induction-deduction method of reasoning
dephlogistated air, see air, oxygen
determinants acc. to Aristotle, 5, 95, 209, 210
dipolar compound, see compound
dipole moment, 192
discontinuity of matter, see corpuscular theory
dissociation degree, 195, 196
  electrolytical, see electrolytical
divisibility of matter, 29
dryness (dry) as principal quality, 32–34, 36, 39, 209
dualistic conception, 31, 38
  theory, 165, 168, 171, 179, 194
dynamics, 209

earth as element, 28, 32–36, 46, 47, 54, 57, 59, 60, 73, 96, 98, 100, 160, 209, 210, 248

effervescence, 48, 70, 135
electric pile, see Voltaic pile
electrochemical cell, 195
　series, 164, 165
　theory of compounds, 132, 164, 165, 168, 171, 194, 202, 248
electrolysis, 152, 166, 253
electrolytical dissociation, 194, 196, 197, 256
electromotive series, 42
electron, 66, 153, 156, 157
　affinity, 198
　as an alloy, 41
　as part of an atom, see atom
　density, 206
　discovery of, 152, 256, 257
　nuclear, 154, 155
　role in bonding, formation, 186, 197, 200, 203–205
　theory of, 152
electronegativity, 166, 259
element (principle), 16, 20, 28–92, 160, 162
　alchemic (spagyric), 41, 45, 46, 51, 55, 56, 58–61, 66, 211, 249
　Aristotelian (peripathetic), 38, 46, 49, 51, 56, 58, 61, 78, 248
　classification (systematization), 121, 134, 135–145
　concept of, 21, 34, 36, 48, 51, 56, 78, 90–92, 118, 211
　desintegration, of, 156
　groups of, 136, 140, 141
　imponderable (fluids), 81, 84, 85, 87, 90
　number of, 57
　symbols of, 113, 122, 163
elementary particle, 91, 157, 158
empiriocriticism, 23, 146
endothermic reaction, 234
energetism, 23, 145, 147, 256
energy, bond, see bond
　concept of, 88, 89, 147, 216, 236–238, 258
　definition of, 209, 211
　equivalence of, 131, 258
　free (Helmholtz function), 238, 239
　internal, 238, 239
　of isomers, 192
enthalpy, 238, 240, 241
　free (Gibbs function), 238, 239, 245
entelechia, 209, 210, 224, 225
entropy, 145, 208, 238, 240, 243, 244, 255
epicycles, 5
equilibrium, 148, 194, 224, 229–232, 234, 235, 240, 243, 244, 255
equivalent, 106, 107, 112, 125, 126, 132, 134, 136–140, 147, 149, 171, 175, 197
　formulae, 126
ether cosmic, 35, 47, 87, 88, 95
exothermic reaction, 235
experiment, 3, 17, 27, 52, 55
experimental science, 18
experimentum crucis, 17

Faraday's laws, see law
fermentation 48, 63, 135
fire, as element, 29, 32–36, 46, 47, 55, 57, 62–64, 66, 82, 95, 98, 208, 210, 211, 215
　its influence on reaction, 19, 48, 51, 55, 65, 66, 72, 98, 210, 212, 214, 215
fluctuation, 148
fluorescence, 152
force, chemical, 226–228, 232, 233
　dead, 220
　definition of, 228
　live, 89, 220
　of the bond, 89
　organizing, 225, 226

gas, as a substance, 48, 54, 63, 69, 85, 87, 89, 105, 109–111, 119, 124, 132, 150, 195, 196, 200, 251, 260
　kinetic theory of the, 134, 145, 255
　origin of the name, 70
　sylvestris, 70
geocentric system (theory) 16, 249
Gibbs function, see enthalpy free
gravimetric (weight) methods, 18, 22, 48, 74

heat, as acting agent, 11, 228
  as principal quality, 31, 32, 34, 36, 39, 57, 209
  as radiant body, 88, 89
  as repulsion sheet, 83, 86, 110, 112, 115, 119, 223
  of reaction, 220, 235–237
  theory of, 81–83, 88, 101, 221, 252
heliocentric system, 16, 250
Helmhotz function, see energy free
hermetic science, 10
holism, holistic approach, 7, 11, 13, 25, 246
homologous series, 172, 178
humors, 29
hypothetic-inductive reasoning method, 17
hybridization, 204, 205

iatrochemists, iatrochemistry, 14, 44, 98, 250
idea, idealization, 4, 17
indicators, 136
induction-deduction reasoning method, 4, 17, 22, 24, 121
intermolecular interactions, see molecular
ion, 166, 194, 202
Ionian philosophers (naturalists), 2, 28, 94
Iron Age, 1, 247
isomers, 168, 181, 185, 191–193, 254
isomerism, cis-trans, 191
  optical, 188, 190, 198
isomorphism, 127
isotopes, 131, 154, 259

kinetics, see chemical kinetics

ladder of buring, 168
  of synthesis, 168
law of combining volumes, 124, 131
  of conservation of matter, (of mass), 3, 18–20, 73, 131
  of constant composition, 107, 125, 149, 252
  of radioactive displacement, 153, 154, 259
  of electrolysis (of Faraday) 152, 166, 194, 253

  of Hess (of thermochemistry), 236
  of mass action, 233, 234, 240, 256
  of mobile equilibrium, 234
  of multiple proportions, 86, 116, 117, 149, 150
  of thermodynamics, 254, 255
Leblanc method, 66, 252

macrocosm, 11
materia prima, primary matter, *prote yle*, **13**, 19, 28–30, 32, 46, 49, 91, 102, **126**, **127**
mathematical (geometrical) approach, 5, **8**, 16, 17, 23, 95, 206, 220, 230, 232
matter (materia), 3, 94, 99
meason, 157
metal, as a substance, 37, 38, 41, 42, 47, **49**, 51, 59, 63, 67, 68, 70, 76, 79, 105, 144, **212**
  base (unripe), 11, 12, 64
  ennobling of, 11, 12, 37, 79
  noble, 11, 12, 37
  origin of, 13, 38–40
  smelting of, 1, 6
mercury, as a metal, 37, 48
  as a spagyric (alchemic) element, 37–40, 45, 47, 55, 57–60, 74, 212, 250
microcosm, (microworld), 11, 183, 204
moisture, moist as principal quality, 32–34, 36, 39, 208
molecular (intermolecular) interactions, 200, 236
molecular constitution, see structure
molecule, conception, 94, 106, 132–134, 139, 145, 166, 171
  constituent, 132
  definition of, 106, 118, 200, 201
  elemental, 132
  origin of the name, 98

names of compounds, see terminology
negaton (negatron), 157
neutron, 91, 154, 157
nitration, 169
nitrobenzene, structure of, 169

non-stoichiometric compounds, see bertholides
nuclear resonance, 193
nucleon, 157
nucleus of an atom (kernel), 91, 201, 205
　discovery, 153, 259
　structure of, 156
nucleus of organic compound, 169, 172–174

objective idealism, 4, 29, 248
Occam rasor, 8
oil as element, 60
optical activity, 187
orbitals, 91, 204
orbits of electrons, 201, 203
osmotic pressure, 195, 196
ouroboros, 13, 14, 185
oxydation, 67, 215
oxydation degrees, (states), 65, 80, 81, 105
oxygen, discovery, 71, 72, 252
　origin of the name, 74
　role in combustion, 63, 77

papyri from Stockholm and Leyden, 6
particles alfa, beta, gamma, 153, 154
passive qualities (elements), 33, 60, 210
periodic system (table of elements), 129, 131, 140–144, 149, 150, 153, 201, 256
peripatetics, 5, 51, 52, 58, 94, 96, 97
philosopher stone, 14
phlogisticated air, see air
phlogiston, 64–69, 73, 76, 77, **216**
phlogiston theory, 18, 48, 63, 65, 67, 71, 84, 145, 252
photon, 23, 147, 154
pion, 157
pneuma (breath), 15, 28
positron, 256
potential, chemical, 240
　thermodynamical, 238
practical arts as chemistry source, 1, 2, 6
primary matter, see *materia prima*
proton, 91, 154

Prout's hypothesis, 127–130, 154
quanta, 23, 156
quantitative methods, 16, 18, 50, 73, 76, 78, 82, 88, 97, 145, 160, 238
quantum, mechanics, quantum methods, 204, 206, 245, 259
　number, 157, 158, 203
quark, 91, 158, 260
quick-silver, see mercury
quint-essence (cosmic ether), 35

radiant bodies, see element, imponderable
radiation, see rays
radical, inorganic, 79, 139, 163, 177
　organic, 170, 173, 174, 176, 178, 253, 259
　theory of, 169
radioactive series, 154
radioactivity, 147
Raman spectra, 192, 259
rate, of nuclear transmutation, 154
　of chemical reaction, 193, 209, 210, 224, 229, 230, 233–235, 244
rational formulae, 179–182, 186
rays alfa, beta, gamma, 153, 156
reaction, chain, 242, 243, 258
　enzymatic, 143
　thermodynamically irreversible, 243, 244, 259
roasting, see calcination

salt, as spagyric element, 45, 57, 60, 160, 212
　conception of, 135, 136, 165
saltpetre, 63, 69, 71, 101
scientific revolution, 16, 246
simple bodies (substances), 11, 79, 90, 162
spagyric science, 10
spatial arrangement of atoms, 145, 182, 185–187, 191, 197, 198, 205
speculum (an alloy), 41, 42
spin, 91, 157, 203
spirit, 15, 16, 28, 57, 60, 69, 70, 135, **216**
stereochemistry, see spatial arrangement
stoichiometric compounds, see daltonides

stoichiometry, 106, 107, 235, 252
stresses, theory of, 190
structure, definition, 180
   of substances (molecules), 133, 159–161, 163, 167, 170, 188
   theory of, 168, 169, 171
sulphonation, 169
sulphur, as spagyric element, 37–40, 45, 47, 55, 59, 61, 63, 64, 66–68, 82, 160, 211, 215, 250
symbols, 13, 14, 47
systematization, see classification
teleology, 5, 18
temperaments, 29
terminology of substances and elements, 73, 161
tetrahedral carbon atom, 188–191
terra fluida (liquid earth), 59
   lapida (rocky earth), 59
   pinguis (fatty earth), 59, 63, 215, 251
texture of matter, 21, 29, 33, 93, 98–102, 106, 149, 161, 162, 213, 215, 251, 258
thermodynamical potential, see potential
thermodynamics, 146, 243, 254
triades, 136, 141
triple bond, see bond

types theory, 169, 171, 173, 175
ultimate particle (atom), 99, 104, 111–114, 116, 118
unicorn, 15
unitarian conceptions, (theories), 30, 169, 172, 255
universities, academies and scientific societies, 9, 21, 249, 251

valency, (atomicity), 91, 135, 141, 144, 175, 177, 179, 182, 183, 188, 196–200, 205
   Abbeg's theory of, 199–201
   ionic (electrovalency), 197, 199, 205
   unit of, 177, 190
vital force (*vis vitalis*), 168, 224, 227, 244, 246
Voltaic pile, 163, 164, 252

water as element, 28, 31–36, 46–48, 54, 60, 96, 98, 100, 208, 210, 248

X rays, 152, 153, 193

yle, see *materia prima*

Zintl's elements, 151
Zeeman effect, 152